Diamonds Everywhere

A Story of Challenges, New Horizons,
Happy Trails, and a Life of Song and Music.
A Journey to Bragg Creek, Alberta, Canada

Siegfried Beckedorf

Paul,

There is Life after 60 !!

All the BEST

B.

◆ FriesenPress

Suite 300 – 990 Fort St
Victoria, BC, V8V 3K2
Canada

www.friesenpress.com

Copyright © 2016 by Siegfried Beckedorf
First Edition — 2016

Diamonds Everywhere is available through the FriesenPress Bookstore
www.friesenpress.com/bookstore

Cataloguing in Publication Data available from Library and Archives Canada

ISBN
978-1-4602-8426-1 (Hardcover)
978-1-4602-8427-8 (Paperback)
978-1-4602-8428-5 (eBook)

1. BIOGRAPHY & AUTOBIOGRAPHY, PERSONAL MEMOIRS

Distributed to the trade by The Ingram Book Company

Table of Contents

♦

TESTIMONIALS

When we moved to Bragg Creek in 1994, the Beckedorf family were already established members of the community.

As neighbours of the Beckedorfs, for many years, we enjoyed the pioneering spirit of Sieg and Ursula, and Lloyd, their son, and Lloyd's wife Linda, of Moose Mountain Log Homes. As a result a log home was designed and built for us on the river. The result shows the "love of wood and trees" evident in the Beckedorf family.

One of the recollections we had was hearing about the wonderful parties that Ursula and Sieg hosted at their home for the senior, the "Snowbirds."

When the Community center burned down in 1998, the Snowbirds were left without a home. Following the fire it was proposed that we construct a new center, with logs. With the support of Sieg, Ursula, and our other members, it was decided to proceed.

True to their love of "wood and trees" they also donated the beautiful Blue Spruce trees from their tree farm, to help with the final landscaping.

True concern and commitment to their friends and neighbours best describes Sief and Ursula Beckedorf. We wish Sieg the best for the future.

—Terry and Gail Graham

As a writer-in residence for the Calgary Public Library I saw many manuscripts but this memoir, written by well-known Bragg Creeker Sieg Beckedorf, made a special impression on me. *Diamonds Everywhere* offers everything from historical facts to international adventure to mediation on life – all delivered with honesty, humour and an obvious love of life and family.

—Barb Howard, 2015/16 President of
the Writers Guild of Alberta

Close friends are truly life's treasures. As also immigrants in the 1950's we feel proud and honoured to have met and know Siegfried Beckedorf, his wonderful wife Uschi and the whole Beckedorf clan living in Canada. Immigrants from Europe in the 1950's brought a lot of expertise and knowhow to Canada and definitely helped to make Canada what it is today.

We met Sig and Uschi in 1965 while I worked at Shell Canada with Sig. They kindly invited us for dinner as newcomers to Calgary, and we have been good friends ever since. We witnessed the pioneer spirit that drove Sig to various levels of success in all his endeavours. And we got to know and love Siegfried as the eternal optimist. Our many get-togethers were always exciting and always accompanied by music and laughter and maybe a Schnapps or two. The greatest gift of life is friendship and we are looking forward to many more years to share it with Sig.

—Dieter and Ursula Cosandier, Cochrane, Alberta

The Eldsons had been in business about thirty years when the Beckedorfs first came to our area in 1961. They came just ahead of some of the conveniences which made living in Bragg Creek less challenging. Electricity was here, but the road, telephone and natural gas were improvements that came after 1961.

In order for the Beckedorfs to access their second property they had to pass our store and gas station – Elsdon's Store / Bragg Creek Trading Post. Since it was one of only two stores in the community at the time it was inevitable that they would meet Jack and Mary Elsdon, daughter Barbara and later Barbara's husband Robb Teghtmeyer. This association has now spanned over five decades and three generations and many facets of interactions.

As Sieg developed new properties in association with his brother Herb, we dug basements, built roads, installed water and septic systems, and many other excavation jobs. It wasn't all work. We attended many rousing parties, Oktoberfests at the Bavarian Inn, the extraordinary costume balls, hosted by the Artisans, and most other community events.

In 1963, on their first weekend property near Bragg Creek, Ursula Beckedorf composed the very touching "Bragg Creek Song", later performed together with guitars with her twin sister Brigitta married to Sieg's brother Herb. This song was very dear to my mother. It was such a special and emotional moment for our family to have the Beckedorf Twins sing that song in the church at my mother's memorial service. It will never be forgotten.

Nor will Lloyd Beckedorf's support and encouragement after the Great Flood of 2013 be forgotten. He and wife Linda arrived at our place the day after with shovels and gave us a glimmer of hope. He inspired us to proceed with the rebuilding and was so generous with his expertise throughout the whole project. It is just another example of how the Beckedorf Family have enriched the community with their involvement in it.

—Barbara and Robb Teghtmeyer, Bragg Creek

♦

PREFACE

"Opa, what were you doing when you were my age?"

This question by Cyr, my then eleven-year-old grandson, and his parents' constant reminders to his grandparents, prompted my wife Ursula and me to put our story together. I am grateful to them for encouraging us to put life into the journals we kept for many years.

I started making notes when I entered the business world in Hamburg, Germany. At that time a friend gave me a notebook to keep a journal. This was very helpful in piecing together things such as encouragements I received (e.g., "diamonds everywhere") and other meaningful events.

Challenging times during and after World War II presented many shocks for a young guy, but also inspirations to develop a positive attitude. Great people led the way, allowing me to look beyond the effects of an insane war to the joy of being.

How can we balance our day-to-day life with the challenges of a world of controversy, violence, and, yet, so much beauty— a world of rapidly advancing science and technology? I can't imagine a day without computers. But can computers shed light on the gift of consciousness?

Ursula and I enjoy travelling and meeting people. While exploring opportunities in business and on private excursions to many countries, I became curious about how intuition and inherent qualities affect our day-to-day life. I often stop in my tracks to "smell the flowers" to remind myself to be open to new thinking and the joy of being.

One author who made me curious to explore new ways of thinking was the East Indian

mystic and teacher Osho. The *Sunday Times* of London named him one of the "1,000 Makers of the Twentieth Century." The world press calls him "misunderstood, blunt, straightforward, funny, while at the same time, brilliant." Visit www.Osho.com.

"From our individual quest for releasing accumulated stress and anxiety due to the most urgent social and political issues facing us today to relaxation into daily life and meaning is only a question of changing our perspective, our approach towards life."

—Osho

♦

MANY SHOCKS FOR A YOUNG GUY

I grew up in a small village that today is called Schneverdingen, south of Hamburg, Germany. Our home was the centre of activities. My grandfather's corner store and post office attracted shoppers on foot, bicycle, horse-drawn cart, horseback, and even donkeys from other villages and towns in the area. I liked the smell in the store of burlap, grains, flowers, and potatoes. At times we helped customers stuff their merchandise into their own bags and load them for pennies in reward.

Our maid supervised our playing outside along a narrow country road, in the forest and ponds. She warned us not to stray to a section of a dark forest where gypsies parked their wagons and camped, laundry hanging in the trees. "They steal chickens and children!" We were afraid of the gypsies and of the students driving their bicycles fast along the country road on their way to school. They chased and scared us like chickens on the run whenever we were near the road. The nearby pond was our favourite playground; summer and winter, there were places for researching crawling critters and bird life.

When I was six years old our family moved to a different surrounding.

We played soccer, hiked into the rolling hills, and rode horses. Singing in the bus on the hour-long drive to high school in Celle was a lot of fun.

In front of my birthplace, circa 1930. There were four children. Elfriede,
Irmgard, Hans, Siegfried, Ewald, Herbert and Edwin to come.

This photo of my birthplace was taken in 1950.

I was nine years old when my mother died suddenly, a few months before World War II began in September 1939. This was the first shock I remember clearly. My mother was a very gentle person, trying to keep five boys from fighting by sending us outside. She was disturbed about the political scene and news in Germany. The clouds of war had thickened. My two older sisters took over the reins and tried to keep the boys in check.

After the start of the war it was mandatory for young people aged nine to fifteen to join the Hitler youth. I was enthusiastic in sports, singing, camping, and marching. I remember singing, "It's a Long Way to Tipperary." We were taught English beginning in grade five.

Initially, the news of victories by German troops made me feel this was good for Germany. We were told that the surrounding countries were envious of the progress made in the re-building of Germany after World War I. We were entertained by the military on Sunday mornings. The public applauded a mounted music band riding through the neighbourhood to the sound of trumpets and drums. I put up posters of air force heroes in my bedroom with thoughts of joining the air force someday.

During the last two years of the war, the mood in our neighbourhood changed. American and British bombers flew over our area, dropping bombs on Hamburg, Berlin, and other cities. Recycling of paper, metal, and cattle bones (for soap making) was organized; food became scarce. We used our bikes to pick up potatoes, rutabaga, cabbage, and buckwheat made available by farmers in the area.

News of German troops retreating on all fronts put fear in our hearts. At the same time, we were told that the German military had developed a secret weapon that would turn the tides of war.

In early 1945, at the age of fifteen, I was trained on a Panzerfaust, a gun-like device used to jump out of trenches then fire quickly at oncoming tanks, disabling the tracks. Others my age, as well as older men, joined me as volunteers. It dawned on me that I might have to kill people. To my great relief, the exercises were called off when our trainers were sent to the front lines of fighting.

Diamonds Everywhere

Three months before the end of the war, I was infected with tetanus, an often fatal disease. Relatives on nearby farms supplied some raw meat. We used a meat grinder to process the meat for different purposes. In the manual grinding process, Herbert, my brother, used the handle while I stuffed pieces of meat into the opening. My ring finger got caught and its tip was partly sliced off. A day later, I got cramps and was rushed to a military hospital since there was no public medical facility nearby. I lay on a stretcher in a very busy place alongside many wounded soldiers and was tied down to control cramps throughout my body. Medical staff stuffed towels into my mouth to stop the bleeding caused by biting my tongue. I dimly remember my surroundings and losing consciousness.

The next morning a doctor told me, "Siegfried, you are very lucky. Tetanus is 99 percent fatal if not treated immediately. We injected thousands of units of medication into your body. Now, you have to listen to the nurses to keep your legs straight; otherwise, you may be in a wheelchair for the rest of your life." I had no choice, as I was still strapped down.

Within a few days, I was able to walk in the hallway of the hospital. Wounded soldiers were brought in on stretchers, some screaming in pain. This was my second shock and was very much imprinted on my mind.

Along with some recovering soldiers, I was sent to a nearby resort for recuperation with my father's consent. The two weeks there gave me ample opportunity to walk with soldiers through the park-like area. Wilfried, a soldier with a seriously injured right knee, only twenty-nine years old, left a positive message with me. Several times during our walks, he referred to the "thousands of diamonds everywhere; there is so much beauty!" Leaning on his cane, he pointed at dewdrops on trees and grass. "I do not want to go back into the trenches," he told me. He encouraged me to look forward to the end of the war and to better times ahead: "Look, *really look*, at the beauty of nature . . . and never lose a sense of humour." Often in years ahead these words made me feel good.

Back home, I noticed many soldiers returning from the front lines, exhausted and depressed, looking very much like the Russian prisoners being transported to camps in the area. The mood in our neighbourhood was one of uncertainty about the outcome of the war. A lot of people expressed their anger and frustration at the politicians responsible for the mess they got Germany into. "If Hitler and his team would have travelled with briefcases instead of tanks, the senseless killing and suffering could have been avoided," a veteran of World War I told me.

Within a few weeks, British military police moved into our town. We threw mud balls at the girls we knew who went out with the officers. We were disgusted and still considered them our enemies. An unexpected notification to move out of our homes to make room for occupation personnel and their families put us into a spin.

All the neighbours got together to find transportation to move our belongings into trucks and horse-drawn vehicles to nearby farms and deserted estates. A large barn-like building served as our temporary accommodation. At that same time, the British freed about three thousand Russian prisoners in the area. We heard stories of rape and murder committed by these former prisoners. The "diamonds" on the trees did not cheer me up. My father and a neighbour left on bicycles to contact the British MPs. Wandering Russians who were former prisoners stopped them and pulled them off their bikes. The neighbour, our former storeowner, was killed and had his ring finger cut off. My father escaped in time to get help from the British. This was another shock I remember well. To the credit of the British MPs, they immediately gathered the marauding Russians in the area and sent them by train and truck east to the oncoming Russians. This quick action by the British was very much appreciated.

After the war, my family re-settled near Schneverdingen, our hometown, on an acreage. We helped to build a small home on a beautiful piece of heather country with birches and ponds. Looking at the beautiful diamonds on the heather was a good feeling.

City of the Hanseaten

My sisters got married and the boys had to look for opportunities to "fly the coop." My preference was to get into the landscaping business, but there were no openings in the foreseeable future.

Robeco, a wholesale firm in pharmaceutical and cosmetic supplies, offered me an opportunity to enter the business world in Hamburg, the "City of the Hanseaten," named after a powerful fleet of northern European cities trading actively for hundreds of years until the nineteenth century. I accepted a position, including a three-year commerce program with the Schlankreye Higher School of Commerce. It was difficult to find accommodation at that time. I travelled by bike and train to downtown Hamburg from my home in the country. Including walking through rubble and ruins, it took about two-and-a-half hours to get to the office in the Chile-Haus by the harbour. Rolf Becker, my boss, and my new friend, Hans Schikkus, a businessman in the same building, praised my efforts.

Travelling with my boss throughout the surrounding countryside to deliver supplies bought from start-up manufacturers in Hamburg was a lot of fun. Rolf was a dedicated businessman with a good sense of humour. A former air force officer, he made good contacts with formerly large corporations like Beiersdorf (Nivea),

Penaten, and Schwarzkopf. These family-controlled firms started out in the basement of their homes and grew quickly. Rolf told me, "Siegfried, we are very lucky to get a shot in the arm by the American Marshall Plan. This will help the German economy get back on its feet."

Robeco was successful and earned a good reputation. On one of our trips we encountered slippery cobblestone roads. A heavy bump brought our old Hanomag truck to a quick stop. The rear wheel came off and travelled past my passenger window and onto the muddy field, where it rolled over. Rolf shouted laughingly in this moment to "get the camera," which we did not have with us. A big truck behind us came to a stop and the driver loaded undamaged cartons of bottles and supplies into his vehicle and followed a tow truck to the nearest service station. We stayed in a hotel to organize the next step. Rolf was in a good mood. Over dinner he talked about having miraculously survived being shot down over German territory as a pilot. He encouraged me to tell him about my experiences during the war. He liked the story about the wounded soldier finding strength and hope in the natural beauty of "diamonds."

The next morning, Rolf was very lucky to secure a new Opel light delivery truck from a nearby dealership. The previous purchaser had not been able to meet all the conditions for the transaction. Future trips were so much more fun with the new truck.

In the meantime, I found accommodation in Hamburg in a downtown apartment building. I enjoyed the Hanseaten entrepreneurial spirit. My friend Hans Schikkus, who lived in the building where our office was located, took me out for lunch many times. Sitting at the harbourfront, watching ocean-going vessels coming and going, he pointed at the ships and said, "It was my plan to sail on one of these boats to America. My leg injury prevented me from following my dream." From him, I think I got some ideas and visualized sailing across the "pond."

Rolf's business was expanding. After about two years he asked me to join him in a move to Bremen, where he was able to take over a larger operation. I decided to stay in Hamburg. My contract with the Higher School of Commerce was taken over by my new employer, Schwitalle & Hesse, a chemical manufacturing company in the same building. I enjoyed the many contacts I made with long-established, family-owned businesses. After I obtained my diploma, I shared my thoughts of taking time off for a couple of years to explore Canada. Hans was enthusiastic about this. "You will not regret that and will come back with a command of the English language and experience overseas. There will be great opportunities in the import and export business. In the meantime, you will miss Hamburg and its entrepreneurial spirit."

I shared these ideas with my brothers Ewald and Herbert. My sister Elfriede, husband Nudung, and their two-year-old daughter Gita had moved to the Okanagan Valley in Western Canada two years earlier. She sent photos of blooming orchards and a huge countryside. I remembered my earlier interest in landscaping and tree nurseries. Looking at these photos while I sat in my office gazing out at the harbour gave me ideas about exploring Canada and the orchard business.

♦

LOOKING FOR ADVENTURE
ACROSS THE POND

Things unfolded fairly quickly. A contact was made in Hamburg with someone who had immigrated to Canada's Okanagan Valley in 1929. Walter Beurich sponsored our trip.

Hapag-Lloyd set up the itinerary: by rail to Rouen, France, via Paris by *Karl Grammerstorf*, a newly built German freighter with cabins for eleven passengers; from Rouen via Swansea, England, to Quebec by Canadian Pacific Railway; from Quebec via Montreal, Winnipeg, Calgary, and Sicamous (BC) on a Greyhound bus to Penticton (BC).

At the beginning of September 1951, we were ready for the trip. My father agreed that a two-year trip to Canada would be a good experience. Still, he was quite emotional when he took us to the railway station, where we also said goodbye to Hans Schikkus. "*Gute Reise* (happy trails). We see you in two years!" It never occurred to me that I could "forget" to come back in this time frame.

A taxi driver was to take us to the La Garde du Nord station in Paris for the train to Rouen. "*Allemagne*?" A taxi driver refused to drive us. I realized it was only six years since the German occupation of France. Rouen was a busy place, the people very friendly and helpful. A strike in the harbour delayed our departure. We met our captain, crew, and other passengers. The cabins looked attractive with a lot of good workmanship in wood finishing.

The choppy waves on the English Channel sent two of the passengers "under deck," the first victims of seasickness. Swansea seemed a sleepy community with little activity in the harbour. We explored life in the town. One sign on a huge church-like building looked interesting: "Dancing from 7 to 9 p.m." Girls, dressed in what looked to me like amended school uniforms, were brought in by bus for sailors to get some entertainment. The dance music was polka and English waltz. Our "Oxford" school English worked well enough to strike up some conversations with girls. By 10 p.m., Swansea was in the dark.

Within two days on the high seas of the Atlantic, strong waves hit the ship. More passengers went under deck. Herr Meier, our steward, told the three of us, after all other passengers were seasick: "This 4,500 BRT (registered tons) ship is well built. Don't worry." We were told to help tie up chairs and tables in the dining room to prevent sliding and "get all the food you can eat." We followed orders. Playing cards with the crew cost us dearly. We had to give up some shirts in lieu of money! Herr Meier had warned us.

Look How Small We Are

Herr Meier became our friend. He was a former chief steward on the German battleship *Gneisenau* and was a philosopher. "Look at the sky, look at the waves. How small we are! In times of heavy seas look inside yourself for strength and courage. I learned to look first at my own self as my ship to steer in the right direction, and then be a good team member doing my part to steer the big ship. The *Gneisenau* was sunk by the British. I was fished out of the Skagerak (North Sea) after we had to jump into the sea. A British officer pulled me out, shouting, "The sun shines on all of us, friend or foe." I made a note of this in my journal, as well as Meier's quotes of Schiller and Goethe. I thought about his wisdom in years to come.

The waves settled down just a few days before we heard a sailor shout, *"Land in Sicht!"* (land in sight). By now all passengers were up front to witness this event. With drink in hand, we wished each other good luck in Canada. *"Das ist Amerika!"* I shouted. I was excited.

"Keiner da" (nobody here) sounds like "Kanada." I was reminded of a joke in Germany, when the first German settlers saw the northern part of the Americas they sent some men ashore who shouted back to the ship, *"Keiner Da!"*

We spent the weekend in Quebec City, admiring the French atmosphere and architecture. The people were very friendly. We met some Germans who had immigrated some time ago. They told us, "Go back to Germany. There is no work here."

Montreal, Winnipeg, and the countryside impressed me: peaceful-looking farms, meadows, and clusters of leaf trees, reminding me of rural areas of northern Germany. The variety of scenes, as well as the riches of this country, impressed me. We were aware of the length of time it takes to cross this big country by rail, but the reality of travelling about four days to our destination sunk in when we crossed the prairies from Manitoba to the Rockies.

We, like many other travelers, could not afford the luxury of dining in the cars provided. We bought our groceries at various rail stations, slept in our seats, and exchanged stories with passengers about the hardships of the early settlers in the Prairies. The land was acquired by cultivating it within a certain time. Homes had to be built despite a lack of timber and roofs were covered with sod. I learned later that many descendants of these settlers were credited with building a strong society in the West.

As we passed through Calgary, I was not excited to see mainly secondhand car lots and few trees. The tallest building was the Palliser Hotel. I changed my mind a couple of years later when I experienced a dynamic city.

The Rocky Mountains overwhelmed me with their beauty and majesty. We changed from rail to a bus in Sicamous to travel to Penticton. The northern Okanagan surprised me with its abundance of lakes. The apple harvest was in the last phases.

Walter Beurich, our sponsor, met us in Penticton for the last leg to Osoyoos in a big Buick car. I respected his ability to speak to us in very good German with a Hamburg accent, considering he had arrived in the Osoyoos area in 1929. He filled us in about his challenges in cultivating his land for growing first a "ground crop," like tomatoes, lettuce, etc. Later, he planted fruit trees when sufficient water became available by "flues" (wooden troughs) from higher locations.

The Beurich couple invited some neighbours in the evening to be introduced to the "new Germans." Most of these settlers had emigrated from different European countries and had not visited their home country since their arrival in Canada. They spoke with a heavy accent and seemed to be hard-working individuals.

The next day, we were introduced to our pickers' cabins, furnished with Okanagan furniture: e.g., apple boxes, makeshift plywood tables, bunk beds, and outhouses. Not much time was wasted before showing us the McIntosh apples to be picked and the twelve – and sixteen-foot ladders to climb, with picking boxes hanging over our shoulders. When I traveled through this orchard area in recent years, I was surprised to see pickers working with six – and eight-foot ladders! The trees are dwarf-like now, to make picking fruit much faster and safer. It was a blessing that we had brought bicycles from Germany, as we were advised. We developed a speedy way of picking apples in different orchards, travelling with our bikes. I reminded myself to pay attention to the "diamonds" on the grass and wild asparagus early in the morning. Before long, we were known as the fastest pickers in the valley.

Walter Beurich right and Siegfried, at Walter's orchard, 1990.

Not much went on in the orchards after the fruit harvest was completed. Job opportunities existed, like plucking turkeys, working in the local cement plant, or digging graves. The local judge assisted us in the search for work. He introduced us to digging graves at ten dollars for a six-foot hole. The sites were marked. This worked out pretty well until a neighbouring hole collapsed and an unexpected visitor appeared.

Two Guys and a Horse

I spotted an ad in the local newspaper: "Man wanted with experience to work in a portable sawmill." Well, I had some experience with horses on my uncle's farm in Germany. A call sent me to board a Greyhound bus to Rock Creek in the Monashee Mountains. A pickup truck was parked at the hotel. One of the two bearded guys asked me, "Are you the guy from Osoyoos?" My bag landed on the back of the beaten-up truck. Without many words, we drove about ten km up a narrow gravel road. The guys stopped halfway to put on snow chains; the snow was getting heavier as we climbed.

In a clearing, I noticed a lean-to plywood shack and some sheds. "Here we are. That's the place." To my relief, a stovepipe was visible on the roof; the ten-by-twelve-foot shed had a door with a large nail to open and a nail inside to close the three-by-six-foot plywood door, a steel bunk bed with an old mattress, and cupboards with cereal and miscellaneous boxes on top. Two Okanagan chairs (apple boxes), and a four-by-eight-foot plywood sheet cut in half resting on apple boxes served as a dining table. A tin wood-burning stove with cut woodpiles next to it and large nails on the wall to hang coats completed the interior design.

Next, I was introduced to "Boy," a big horse who looked curiously at me. Some oat bags, bales of hay, horse gear, and a metal tongue for picking up logs were all under a large lean-to shed. Louie, the older and more talkative guy, showed me how to use the steel tongue and how to throw the reins over the horse's back once the horse was geared up.

"He knows his way back and forth from the pile of logs cut to the mill," Louie told me. The portable sawmill was located a little down the hill. The mill consisted of a big circular saw on tracks mounted on two large timbers.

The next morning, I woke up in a very cold place; I knew I had to put larger and possibly knotty pieces of wood into the stove to retain some heat. As I walked around my palace, I noticed bear tracks in the snow, possibly grizzly. Well, I knew not to leave any food items outside! Boy greeted me with a friendly call. I fed him and he patiently endured the gearing-up procedure. I enjoyed the work in sunshine and glistening snow: diamonds everywhere. The pile of logs to be cut to ties grew by the day. I kept a timesheet and had it initialed every day. My two employers proved to be very unreliable. The mill broke down often. This made it necessary for them to drive to Rock Creek for parts. Although I provided a list of groceries I required, the list was never completely filled. After two-and-a-half weeks of not getting paid, although they had promised to pay weekly, I became suspicious. I considered riding the horse to Rock Creek to sell it to a rancher in compensation for my wages. One day, after a load was being readied for hauling to the CPR in Rock Creek, I threw my bag onto the truck and told Louie I had to get some supplies in town. He grumbled but that did not stop me from jumping into the truck with them.

An RCMP car was parked near the railway station across from the hotel. I took my bag and walked over to a nearby officer and explained my situation. "These guys are trouble," he said. "I will see to it that you get paid." After picking up their cheque, the two men came to the hotel to cash it. The RCMP officer joined me in demanding my money before they paid their hotel bill.

I was paid in full and thanked the officer with a firm hand-shake. He wished me well. I walked over to the two men and said, "The sun shines on all of us, friend or foe. I wish you well and goodbye." Somehow, these words came to me, remembering Herr Meier's wisdom. I took the Greyhound bus back to Osoyoos.

Winter in the Okanagan

Back in Osoyoos, I looked forward to learning how to prune fruit trees. By the end of January, I was ready to take on contracts to prune first apple trees, followed by cherries, peaches, and apricots. It was a profitable undertaking and kept me busy until the end of March, with many opportunities to rest from climbing ladders in order to look at "diamonds" on the wild asparagus in the orchards.

In early January 1952, snow was piling up. My sister Elfriede had to be rushed to the bus station on the highway to get to the hospital in Oliver. Nudung, my brother-in-law, organized a tractor and trailer with an upholstered chair to transport her. What a sight! A happy girl, Alice, arrived just in time.

A year later, my sister Irmgard, her husband Hermann, and my youngest brother Edwin arrived in Canada from Germany. Edwin had no problem adjusting to the new conditions. He graduated from high school and entered university in Montreal after a short stay in Calgary.

I took my sister Irmgard to the Oliver hospital to give birth to Wolfgang in February 1953. Our families were growing by leaps and bounds. We seemed to be spreading our wings.

In the spring of 1953, Walter Beurich offered me an opportunity to buy his orchard over time. I was tempted, but declined. I discussed this with my brothers. Ewald, Herbert, and I felt it was time to move on from the orchard business to check out sawmills. We learned of constant strikes in many of the larger sawmills in the interior of BC. Jobs were available at the Ponderosa Sawmill at Monte Lake, near Kamloops. While Ewald and Herbert were placed at the "green chain," meaning pulling cut lumber off the conveyor, I was asked to guide logs on the lake toward the mill. Those logs were taken off logging trucks coming in from the surrounding hills. Balancing myself on the floating logs took some practice. This went well until I hit a long pole with a hook (the guiding tool) into my toe instead of a log. I fell into the water and was quickly pulled out by nearby workers. My boot showed blood so the boot was cut off and a bandage was placed on my foot. A van took me to the hospital in Kamloops. A two-week "holiday" in the hospital allowed me to purchase a Remington typewriter and use the time to type my assignments for the business correspondence course I had started in Osoyoos.

Back at the sawmill in Monte Lake, I was placed next to a Japanese fellow to assist in the cutting of two-by-four lumber. We had to communicate in sign language as his English was not good. He tried hard to explain to me that his daughter was teaching square dancing in the evenings and on weekends. We enjoyed those lessons by his daughter, who was very beautiful and spoke English well.

Several immigrants working at the sawmill joined us on a trip to Kamloops to go shopping and have a beer. We walked into a beer parlour and sat down at a large round table. We noticed a group of young men sitting close by. Their table was covered in beer glasses and they were talking very loudly. Shortly after, five Indians came in and sat down close to our table. We noticed the "loud guys" looking bewildered at the group of Natives. We knew that in BC, Natives had only been allowed to go into public drinking places since earlier that year. We were quite disturbed when one of the lumberjacks, as the loggers in the mountains were called, stood up and shouted some obscene words at the Natives. As he moved toward them, a tall Norwegian in our group stood up, as well. "We can't let these guys behave like that" were his words to us. We stood up, too. The bar owner now got involved. "Pay up and get the hell out of here," he told the group. They obliged hesitantly. We joined the Natives in this confrontation to show our support. They and the bar owner thanked us.

Several strikes took place at the mill. We had to join the union and were promised strike pay. The job steward explained that we were entitled to strike pay only after six months on the job. This was contrary to our understanding. We were fed up with this business and demanded our pay. The office advised us that the cheque would be mailed to a Calgary hotel. We did receive the money, in time. It was our intention to drive to Calgary on the way to the oil fields around Peace River, Alberta.

To the Oilfields via Calgary

Alberta's Peace River country offered a lot of opportunities in the oil business, but a stopover in Calgary changed our plans. A sign at the western outskirts of the city read: "Calgary, the Stampede City, population 157,000." How exciting to arrive at the beginning of the Calgary Stampede, the "Greatest Outdoor Show on Earth!" Horses, cowboys, cowgirls, and Indians in original attire; BBQs everywhere, pancakes downtown and in the suburbs!

What energy on display! We decided to stay around and check out this so-called cowboy town. Hotel Noble was recommended as a reasonably priced place within walking distance of all the exciting events in downtown Calgary.

Square dancing in the streets of Calgary—what a sight! We were walking along 7th Avenue when a group of square dancers opened up their line to invite passersby to join them. I did not hesitate to fall in step. A group of Western riders swinging their hats followed by colourful bands playing western music provided an atmosphere of the Wild West.

We visited the Rocky Mountains and hiked up to Lake Agnes and the Beehive Trails above Lake Louise. We made friends in the city and enjoyed life in Calgary. I experienced the spirit of this dynamic city and the phenomenal growth in population, business, and merging of cultures, all the while keeping alive the Western spirit. Later, when I introduced German investors to Calgary as the LRT (Light Rail Transit) was being constructed above the city core, a German real estate professional commented, "Now is the time for city government to deal with public transportation, tunnels underneath downtown, and attention to density around the downtown core rather than sprawling suburbs."

"You Are a Good Dancer"

The Polish Canadian Community Hall was known to immigrants as a place to go dancing and have fun. We checked it out in very casual clothing, looking more like Hawaiians. I noticed a group of young ladies assembled in the corner of the hall. I approached one of them with her back toward me facing the others in the group. Tapping her on her shoulder and asking, "May I have this dance?" resulted in substantial changes in my life.

The surprised look on her face embarrassed me. What a dancer—so light, so lively! I learned that Ursula had arrived a month before from Germany with her five-year-old daughter, Monika. Brigitte, Ursula's twin sister, was also in the group. She had come to Calgary from Germany six months earlier. I had some difficulty telling them apart, as they were dressed in a similar fashion. The band was good and lively, playing polkas, swing, fox trot, and beautiful Viennese waltzes (Walzer).Ursula sang along with the live band while we were dancing. I was in awe of her beautiful voice!

We met often after this dance and made many trips into the mountains and foothills. Ursula had divorced her husband and was able to flee from East Germany to West Germany with Brigitta. Ursula told me about her adventures smuggling Monika across the border between East and West Germany during the night in a baby buggy. We visited the forested and hilly area later, as well Sonneberg, where the twins spent many years after their return from Africa. The watchtowers erected by the East German government to prevent crossing the border were still in place. Ursula had been lucky to escape. The consequences of having being caught would have been quite serious.

Monika attended elementary school. I was surprised how quickly she began to speak English. I enjoyed our trips with Monika into the mountains and visiting the Calgary Zoo. On one of our walks in the neighbourhood, I showed off my skill of walking on my hands and playing soccer with small rocks. She was quite impressed. We became friends.

As Ursula's and my relationship became more serious, I decided to join my brother Herbert traveling to Montreal by train. I needed time to think about a future with Ursula and Monika. Brigitte and Herbert were dating and Herbert started evening classes for a bachelor of commerce program in Montreal. It did not take very long for me to return to the West. I missed Ursula. Any thoughts of returning to Germany soon were blown away by the "Four Strong Winds" of Alberta.

I found an interesting position in the food industry as an assistant office manager at Jenkins Groceteria, a wholesale operation. Ursula joined the Arthur Murray School of Dance. This was an opportunity for me to become a better dancer. We met often in the Pig & Whistle restaurant on 8th Avenue, below the dance academy. It was here that I put a ring on her finger— our engagement!

The winter of 1953-54 was one of the coldest on record. Sometimes at a bus stop—cold feet, cold ears—I wondered whether it was a good idea to settle in this frigid climate. We endured the cold and began to appreciate the clear and very blue skies in Calgary in the wintertime, as well as all year round.

We made plans to get married. A small house with a large garden became available for rent. Turning over the soil for growing potatoes and other vegetables was good exercise.

We set the date of May 21, 1955, for the wedding. Our growing family and friend circle joined us. Monika, in the middle of the crowd, told them, "We are getting married!" The diamonds on the grass seemed to direct their sparkle at me all day long, setting the theme for a good time! The next day we went to Banff for our honeymoon.

While Ursula and Brigitte found many opportunities to sing and play guitar, I assisted them in getting to the locations for their performances. Sometimes, Monika, then about seven years old, came along. Noticing my secondary role at these performances, she put me at ease by telling me, "Dad, all you need is a little more practice singing." Even though I tried to follow her advice for about sixty years, only the birds in the forest appreciate my singing.

Ursula and Siegfried's wedding, 1955.

A GROWING FAMILY: NEW HORIZONS

"Omi's Wisdom"

In 1955, we welcomed the twins' mother, Kaethe Wiegand, "Omi," and their sister Barbara, from Germany, to join us in Calgary. Omi had had a stroke in 1943. She was paralyzed and had lost part of her speech and movement on the right side of her body. The twins' wish to bring her to Canada was repaid by Omi in many ways. Her positive outlook on life in spite of her limitations inspired all of us.

The arrival of Omi, 1955.

She brought books and magazines on "mental positivism" with her from Germany. I devoted a lot of time to studying ancient wisdom as living knowledge—a key to understanding the meaning of life. I printed a copy of a poem Omi kept framed by her bedside.

In jedes Menschen tiefstem Grund Tut
sich sein wahres Wesen kund Erschliesst
er diesen klaren Quell, Erstrahlt sein
Leben froh und hell.

I translated this as follows:

Deep within us
We find our true self.
Once truly conscious of this source,
Our lives will be filled with joy and happiness.

Much later, I rewrote this poem in my own words:

Out of my deepest self—
I sense being part of nature, its energy, its mysteries.
With conscious breathing, I direct this energy
into my own power
and the joy of being.

"Who looks outside, dreams; who looks inside, awakens."

—Carl Jung

Moving On

A bungalow in Crescent Heights became our new home for our larger family. The Irish owner, Mr. Quinn, was very helpful in our search. He arranged a suitable way for us to purchase our first home. Ursula maintaining the big, beautiful weeping birch tree in front of the house was my motivation to make the deal. Mr. Quinn told us, "This birch tree needs to be watered a lot before frost comes; otherwise, in the spring, it will get damaged by our Chinook winds. The sap will rise and the tree will lose its tip by late frost in the spring." I later passed on this advice to a future buyer of the property.

Lloyd with Grandfather, Opa, 1959.

A German immigrant and carpenter built a suite in the base-ment within a year; in return, he rented the suite. Both our chil-dren, Lloyd and Korina, were born in this house. Our Omi enjoyed looking after her grandchildren. What a nice picture seeing her sitting next to the babies, feeding and playing with them on the porch in the shade of the weeping birch. My father visited from Germany to meet the growing family; he was quite emotional and curious about life in Canada. He and Monika took long walks with Lloyd on their hands. He took Korina in his arms and commented, "To me, she looks just like your mother." He had tears in his eyes. The big family, including my father, visited the Rocky Mountains. He was well acquainted with the Alpine country in Europe; the Rocky Mountains looked to him rough and jagged. "A sign of them being much younger," he commented.

Our growing family, 1963.

One-and-a-half years after our wedding we had a double wedding under the weeping birch in a colourful fall setting. My brother Ewald and Lilo, as well as Brigitte, Ursula's twin sister, and my brother Herbert, got married. As I prepared the site for the wedding in the morning, I was greeted by a brilliant glitter on the moist grass: diamonds everywhere. What a reminder to look forward to happy days ahead! The following day we travelled into the mountains, laughing and singing German folksongs all the way.

Double wedding, 1956.

Walking on My Hands

Out of the blue, once in a while, for as long I can remember, I had an urge to walk on my

hands. I carried this on until Ursula banned my athletic efforts on my sixtieth birthday—bones becoming brittle; I could get hurt.

I remember after moving into a new house in Calgary, I filled my pockets with change. Whatever coins fell out when I was upside down, the children could pick up. Around open fires, along sidewalks and trails, whenever I felt like it, I put my feet up into the air. It was an expression of feeling good.

top-left: Siegfried, age thirty-two, 1960; top-right: in Germany, age fifty-three, 1982; bottom: on sixtieth birthday, 1989.

Our children, nephews, and grandchildren tried to join in, lifting only their heels. For some reason, other adults did not follow my exercise. Diving off a 10-metre tower at the swimming pool in Happy Valley (now Valley Ridge) did not encourage others to follow, either.

On my birthday, while with friends in Wuppertal, Germany, I could not resist the urge to walk on my hands in their living room. I got a photo later showing my handstand holding up the ceiling. The house was built in the eighteenth century with low ceilings.

San Francisco, Here We Come

Brigitte and Herbert returned to Calgary after seven years in Montreal. We celebrated Herb's accomplishment in earning a bachelor of commerce degree.

Our two families enjoyed camping in the foothills and mountain parks as far as Vancouver Island. The idea to camp all the way to San Francisco became a reality while we were still living in Calgary. We updated our camping equipment with new tents and stoves and headed out in May, leaving when there was still snow on the ground.

Setting up and disassembling camp at the first two camping places—one in Yak, BC, another at Moses Lake in Washington state—was a lot of hard work due to heavy rain. The children, whenever possible, helped (Kevin, then the youngest of Brigitte's and Herbert's family, was still in diapers). The most difficult part was handling the big canvas tent because of its weight when wet. Making a campfire was also a great challenge.

What a relief when we camped in Ellensburg, Washington. The sun came out. Spring had sprung everywhere. Fruit trees were blooming. The children found their shorts and ran in all directions. By this time, a routine had been established and all of us had certain roles to play.

The drive along the Pacific in Oregon overwhelmed us. The junipers, the rhododendrons, and all sorts of flowering shrubs framed the coastline in a most beautiful way. Camping at Standish Hickey National State Park in California offered the opportunity to explore the beautiful mixed forest of pines, walnut trees, and jungle-like underbrush. Lloyd, in shorts, got into trouble with poison ivy. Korina and Ron tried to climb trees and got a lot of scratches. The twins were happy and relaxed. I took a deep breath full of the smells of the variety of growth, the diamonds on the shrubs, and the sun shining through the huge acacia trees. Everything was so refreshing.

Camping near San Francisco, 1963.

The foggy, cool weather in San Francisco made us huddle by the oceanfront. The children shouted and pointed at a man coming out of the water in fishing gear with his hood up. I called in excitement, "This man probably came from Japan walking on the ocean floor." After a while, I was looked at in disbelief.

On the way back, Ursula had a very painful toothache. She could not sleep so resorted to wine to overcome the pain. The next day, a Sunday, we located a dentist who told us how to get to his office. He took care of it.

The camping duties were well shared to allow time to sit and sing around open fires. Oh, what fun! We returned via a different route and enjoyed the scenery. The brown grass and leafless trees close to the Canadian border were not inviting at all! Brigitte was relieved when we arrived in Calgary: Kevin was out of diapers.

Looking back, I consider this trip significant for both families. It was the first for the two families since Herbert and Brigitte had returned from Montreal. The children seemed to be closer to each other.

Before moving to Bragg Creek, the twins, known by then as the Beckedorf Twins, entertained at clubhouses, the Calgary Stampede and also with Wilmar Lar and his Irish Rovers.

THE FOREST CALLS

"Trees are poems that the earth writes upon the sky."
—Kahlil Gibran

The forest calls, 1970.

At times, I have been called a "tree nut." My interest in all species of trees started at an early age. I grew up on the edge of a forest of oaks, alders, beeches, birches, pines, and spruce. I remember climbing beech and oak trees was a lot of fun for my brothers, friends, and myself.

During my travels in various countries I've had many opportunities to see diverse and remarkable species of trees. In 1992, I picked up a book in Germany called *Kapital Wald: (Capital Forest) Eine oekologische Bestandaufnahme in Bildern*, an ecological inventory in pictures, covering all aspects of forest management with detailed descriptions of the history of trees species in Germany, Austria, and Switzerland. I carried this book with me while travelling through Germany and Eastern Europe. Many Germans are tree lovers; I've discussed numerous s issues mentioned in this book with foresters and landowners.

1,001 Trees in a Beetle

We planted forty trees on our pie-shaped property in Thorncliffe, Calgary. The clay-type soil required loam and water. A neighbour recommended I pick up some moss in the Bragg Creek area to keep the soil moist. We loaded up many boxes around the Watson Sawmill, west of the provincial park. Looking at the tall spruce trees around me, I said to my family: "Such a beautiful forest and so close to Calgary, a prairie city." The idea of looking for a weekend place here made sense.

In 1963, we purchased an acreage for weekends near Bragg Creek. A cabin on the property served our needs while Brigitte and Herbert built a Lindal cabin, the first Lindal project in Alberta. We spent many weekends on the land of pines, spruce, aspen, willows, and some meadows. The children named trails after the predominant wild life present—i.e., rabbit, red fox, deer, and grouse. Ursula spent some time by herself on the property playing the guitar and writing and translating German and English folksongs.

My tree-planting passion was fulfilled when we purchased a thousand spruce seedlings and one small weeping birch. The transport in a VW Beetle became a challenge with two children, a boxer dog, and two adults! There was a lot of moisture in April, making the ground suitable for planting these seedlings. Merely using a spade to open the ground and sticking the seedlings into the clay type of soil accomplished our goal. Due to lack of water, the trees grew slowly in comparison to the two thousand blue spruce trees we planted later on in the Bragg Creek area.

Barbara, the twins' sister who lived in Toronto, once visited us during cold winter days. In a crowded cabin, we managed to stay warm by means of an oil heater. This was quite an adventure for Barbara, who had no camping experience. We played games while Lloyd and Korina were already sound asleep in their sleeping bags. We had some difficulty getting our VW started in the morning. Placing a heater underneath the oil pan solved that problem.

The three sisters in Lake Louise, 1965.

Too Many Sticks for Ranching

After three years, we sold our first acreage: a proposed subdivision of twenty acres into two ten-acre lots did not meet municipal regulations. We bought an acreage closer to Bragg Creek. We acquired this second "Beck 'n' Busch (Beckedorf and Buschendorf, the twins' maiden name)" acreage of forty acres from Cliff Thoroughgood, the founder of Totem Building Supplies. Cliff told us he wanted to retire in Victoria. Two years later, I met him again. "Retirement is not for me," he told me. He started Totem Building Supplies Ltd. and was very successful.

The access road needed improvements. Gene Fullerton, the local Texaco service station owner, recommended we contact the road builder involved in widening the road bypassing this acreage. "Give the boss a couple of bottles of rum and he will improve your road on a Friday evening." That suggestion worked; the road up the hill was widened by about midnight one Friday evening. We had to better the treat for the crew.

Again, I planted spruce trees along the driveway. The land offered a beautiful view of the mountains and the Millarville hills. Finding a supply of drinking water was a challenge. A clam-shell type of backhoe dug the existing watering pond for cattle to a depth of
more than three-and-a-half metres. That yielded lots of water, but not of drinking quality. About twenty years later, Lloyd and Linda bought the adjacent large acreage and built their dream log home. Looking down today from that site to the acreage we bought in 1967, I see the tall blue spruce trees, the traces we left.

Our land was situated across from the homestead of Stanley Fullerton. He was one of the first settlers in the Bragg Creek area in the early 1900s. Stanley lived west of Calgary, in a subdivision now called Discovery Ridge, with his parents. He rode his horse, sometimes with a wagon, while he cultivated his homestead to obtain title. His family settled down on that land a few years later. "Grandma Fullerton," his wife, maintained a big vegetable garden for the growing family.

We visited Stanley often and learned a lot about the history of Bragg Creek and area. He told us that the well-known ice caves once allowed only a few feet to walk in because of the presence of solid ice. "Now, you can walk in about a hundred feet. The ice is retreating." We confirmed this later when Herbert and I rode to the foot of Prairie Mountain, cobbled the horses, and walked up to the ice caves.

During one of our visits, Stanley asked, "Do you want to buy more land?" When we looked surprised, he said, "My sons are selling some land in West Bragg Creek." When I asked him why they wanted to sell, he replied, "Too many sticks for ranching," meaning too many trees. I replied, "I like trees." He encouraged us to "go see them." Those three words were like a lightning rod! It changed our way of life.

A visit with Bob and Jim Fullerton and Alex Dahrouge made us very curious. In my mind's eye, I saw tall conifers, meadows, and valleys. "Here is a map. Go and walk around."

Herbert and I had hiked in many areas of South and West Bragg Creek, but this stretch of land was overwhelming. We followed old wagon roads and game trails. Some old barbed wire fences were strung below the highest elevation, torn down in places, and no survey lines! "We need to get rid of those fences!" I said. We got lost in the forest.

Bob Fullerton showed us around so that we could get a better idea of the land. We asked for time to talk in more detail. The following weekend, Herbert and I carried a picnic table up to a clearing in the forest. The families followed in the afternoon for a picnic on the land. What a mountain view; what excitement! The children became adventurous. We had to watch them so they did not get lost.

Most of this piece of land was covered in a forest of spruce, pines, aspen, and muskeg, stretching along Kananaskis Provincial Park on the western and southern side, to the east on Fullerton Ranches land, and north to a narrow road access to the Kananaskis. In total, 130 hectares; the terms thirty days cash! We were overwhelmed.

We put our heads together, walked the land a few more times, and got excited about this opportunity. How to pay for it? What to do with so much land? I approached the main branch of the

Royal Bank of Canada in Calgary. "We don't fund raw land—besides, do you expect people to live out there?"

We were given more time. I suggested we get a surveyor first of all. Herbert and I helped to cut survey lines.

After leaving my position with Shell Canada in Calgary, I was looking for opportunities to become independent of big corporate life. I obtained a real estate license and advertised an acreage in the Priddis area for sale, southwest of Calgary. Drs. Michael and Priscilla Barnes replied to my ad. The acreage did not quite meet their needs. Michael said, "We need a peaceful place in our demanding professions. We are looking for an acreage for weekend purposes." Michael was the chief medical officer at the Foothills Hospital and Priscilla was a corporate medical doctor. I mentioned our opportunity to buy land in West Bragg Creek. "Let us see it," Priscilla asked. We walked up the new survey line. Michael asked me, "How much land is this, from the tall spruce tree there and here?" I had no idea but told him I would find out and let them know as soon as possible. We came to terms and the Barneses built a home and still own their acreage today.

We proceeded with the purchase of the land. Bernard Laven of Laven & Laven was our legal expert. After looking over the agreement, he said, "This seems to be an excellent investment. Why don't you 'sit on it' for a while and let the value increase from year to year?" In retrospect, this may have been a good idea! But our love of the foothills and plan to live on the land, as well as the commitment to sell a parcel to the Barneses, made us proceed with the subdivision. We realized this would mean a major change for our children but, at the same time, possibly a good experience.

Looking back, Bernie Laven's suggestion to "sit on it" deserved a little more thought at that time. We faced a time restraint when purchasing in 1969 and it was a lot of land. The excitement of being able to live in the country grew by the day.

To meet the financial conditions to which we committed in Drs. Barnes' Offer to Purchase, we were required to subdivide the land. There was an option to carve out the required land for the Barneses but delay our move to the country so as to allow us to adjust to such a major change in our lives. This thought occupied our minds as we experienced the pros and cons of meeting challenging municipal and provincial regulations. My preoccupation with the processes of subdivision over a period of a few years diverted time from enjoying the outdoors. At various times, I felt like holding off the road builders when trees came crashing down and dirt was being moved around. I was frustrated. On one occasion while contractors were building the access road to the parcel sold to the Barneses, I walked in front of the Caterpillar to stop it from getting too close to trees that were not necessary to take down. I was looking for peace and quiet times in the forest and more time to spend with my family while sharing the beauty of the country. I consoled myself with a plan to plant blue spruce trees along the easement to be cut into the forest. While the subdivision proceeded, we planted two thousand blue spruce seedlings.

Standing in the middle of this beautiful forest, I was aware of the stillness, the chaos of untouched forest, and at the same time a natural order of nature. Some trees uprooted by natural causes had new life, thriving out of rotten and decaying matter.

I was conscious of being part of this great natural order of birth and death. What a gift to experience consciousness! Who am I in this natural experience? The atoms that make up my body were once forged inside stars and, as the forest around me, connected to the whole in incomprehensible ways.

There is power in this consciousness, an awareness of my own power to look at life, at times detached from the busyness of the world around me. I felt a respect for new life in the chaos, yet sensed the harmony and order of the forest.

Those Back in the Bush

We named our new acquisition "Beck 'n' Busch No. 3." Our neighbours were quick to call us "those back in the bush."

We had a vision of building a road to follow the contour of the land. When I presented this to the council of the MD Municipality of Rocky View, with reference to the winding roads in Alpine countries in Europe, I was told, "We are not in Europe. We require straight roads for our school buses!" We appealed this decision, which caused a delay. Considering that today the straight road has become almost a racetrack for motorbikes, I believe the MD now regrets its decision.

First house Beck 'N' Busch No. 3, 1971.

Beck 'N' Busch No. 3, 1971.

I spent a lot of time with MD of Rocky View staff, surveyors, well drillers, and road builders to follow through with our plans. It was also a time I spent in the stillness of the forest, refreshing my mind with encouragement and "wisdom of the ages." Outdoors: open skies, an open mind, an expanding consciousness. I was reminded of a song, "Don't Fence Me In." All this added to my enjoyment of the almost untouched wilderness. A few meaningful words from Nietzsche came to mind: "For happiness, how little sufficeth for happiness! . . . The least thing precisely, the gentlest thing, the lightest thing, a lizard's rustling, a breath, a whisk, an eye-glance—little maketh up the best happiness. Hush!" I added: "And look into a deer's wonderful eyes."

The twins followed a dream they had—an idea to immigrate to Canada to live in a log cabin, get a shepherd dog and horses, and look after children in need. Our plans to build a log home on the new land did not materialize; our time frame was too tight. Two Dutch carpenters offered to build our European-style frame home. They put up a tent and enjoyed the work "in the hills." Lloyd and Korina attended school in Springbank. There was only one school bus, driven by Jack Merrifield of Bragg Creek. This was a major change in routine and took a while to get used to.

We built a corral for our horses. Bob Ecklin of Elkana Ranch provided the children with training in Western riding. A sad accident hit us all very hard. Black bears scared our horses and in the process a filly was killed by a broken rail. Korina, who had developed many skills in looking after the horses, was especially heartbroken. Her horse, Darledo, was often confronted by her dachshund, Buffi, who stood up against the horse after she was in the saddle! Raising dachshunds (dackels) as well as Siberian Huskies was a lot of work and fun, especially for Ursula when Husky puppies were born under our porch in the winter with only lightbulbs for warmth. Lloyd's main interest was in exploring the wilderness. On March 7, a cold, snowy day, Lloyd decided to go backpacking into the Prairie Mountain area. He was fifteen at the time. Ursula took him along the Elbow Falls Road and, with mixed feelings, saw him walk off into the snow-covered hills. Within a couple of days, he decided to walk back to the road and got a ride home. This experience, I know, did not deter his love for the foothills.

Korina, with filly, and Siegfried at Beck 'N' Busch No. 3, 1972.

In the meantime, Monika travelled through Germany, Italy, and Spain. She entered the University of Edmonton and eventually moved to the Vancouver area. She set up a greenhouse operation and tree nursery. We had many opportunities to spend time there and help Monika in her greenhouse operation.

The twins got involved in the (at the time) very small Bragg Creek community. Ursula served as president for a term and was a founding member of the Bragg Creek Performing Arts, the Ladies Auxiliary, and the Bragg Creek Singers.

URSULA BECKEDORF

Commencing from the moment she became a resident of Bragg Creek, Ursula Beckedorf has devoted herself to the enrichment of life in the hamlet. For more than twenty years she has been involved in many aspects of volunteer service; her unstinting dedication is an inspiration to the entire community.

Ursula is in her second year as President of the Ladies Auxiliary and has served with distinction on the Board of Directors of the Community Association. Beyond such elected positions, her contributions comprise an impressively wide-ranging list being the Bragg Creek Snow Birds, Seniors Fellowship, the Bragg Creek Ladies Auxiliary and has volunteered countless hours in the Foothills Hospital's Alzheimers Unit.

In the seventies, Ursula was instrumental in creation of the Bragg Creek Artisans. In conjunction, she then organized the Artisan Singers; this popular group, now numbering 17, has brought pleasure throughout Southern Alberta with its generous entertainment for senior citizens in lodges and hospitals. Performances are always closed with the beloved "Bragg Creek Song", composed by Ursula.

Ursula brought the first "Performing Arts" program to Bragg Creek (through "Alberta Culture") and performances continue to be sold out each season. This success must be largely credited to Ursula's exhaustive efforts in procuring fine artists, fund-raising, phoning, selling tickets and arranging publicity.

In the eighties, Ursula organized the live music for the community dances, raising needed funds for the new Hall. She has also instigated the enjoyable "Oktoberfest" evenings and managed the preparation of memorable German feasts. Her other fund-raising activities include involvement in the annual Bragg Creek Parade and the Beer Garden events.

Ursula's rare energy, enthusiasm, organizational skills and hard work have enhanced the quality of life in Bragg Creek immeasurably. Her selfless devotion and sparkling personality have endeared her permanently to this community.

Trips into the Rocky Mountains were always a highlight for our "Omi." We left her in the lobby of the Chateau Lake Louise. With binoculars, she followed the families' hikes up to the "Beehive" and to Lake Agnes.

Daily commuting to downtown Calgary for Herbert and myself was initially quite a challenge. We acquired a snow plough to keep our roads open. Improvements to the West Bragg Road took a long time to meet our needs.

The population in our area increased, with families joining us to live "in the sticks." Barbecues with the twins playing guitar and singing around open fires, potluck suppers, and games of bridge were very welcoming to our new neighbours.

TUESDAY, JUNE 13, 1972   ROCKY VIEW NEWS AND MA·

Delightful duo

Ent
need
sects
then
met,
rec
biolo
and
inse
iner
supp
C
biolo
requ
pest
for
are
the
dev
occu
san,
inse

BRAGG CREEK TWINS Ursula and Brigitte Beckendorf serenaded the literary elite exti
of Calgary and district at an authors' banquet held Friday. Pictured with them is their che
mother Mrs. Wiegand.
                mor
        (Photo by Ed Arrol) whic

The twins, 1988.

Our log home, 1980.

Family reunion at log home, 1982.

The gardener with dachshund, 1984.

Korina and Darledo, 1980.

Paul Beckedorf assisting in tree nursery, 1983.

Beck 'N' Busch No. 4, 1979.

Raising Huskies, 1982.

Family reunion, with the seven Beckedorfs, 1993.

Our families' happy times in the foothills were overshadowed by the untimely death of Mark, Brigitte and Herbert's youngest son. Mark was a happy young man venturing into the international world of the oil business. He died in 1996 after a challenging period of cancer. A few years later, while visiting Brigitte and Herbert in Yuma, Arizona, Ursula and I shared in the sad news of learning that Ron, their oldest son, had passed away in Kelowna. Paul, Ron's son, was deeply disturbed about the sudden death of his father when he telephoned from Kelowna to let us know. These were difficult times for both our families.

♦

MIXING BUSINESS
AND PLEASURE

An Invitation to Munich

In 1974, I met with a group of German investors who were looking for opportunities in the oil business. A tax-saving agreement among Ottawa, Washington, and Bonn was the basis for this group to open an office in Calgary. After a meeting in Calgary, Ursula and I were invited to Munich to meet with the principals.

This first trip back to a New Germany, after we had left Germany still in ruins, was a wonderful surprise. Gleaming new airports in Frankfurt, Hamburg, Berlin, and Munich; newly landscaped autobahns; new homes, apartments, and office buildings! Our accommodation was in the Deutsches Haus, a hotel dating back to the eighteenth century. "Loden" uniformed staff greeted us. When they bowed down to welcome us, I asked Ursula, "What are they looking for?" The suite included a large bathroom with an old-fashioned bathtub. I was curious and pulled on the golden chain hanging off the wall. This brought a very concerned staff member to the door. "Are you OK?"

While I attended business meetings in Munich, Ursula received a tour of the city. The headquarters of the investment group were located in buildings dating back to the eighteenth century, with high ceilings and big windows restored to their original look.

I was impressed by the courtesy, efficiency, and optimistic outlook displayed by the principals. The plan was for me to open an office in Calgary to establish contacts in the oil business with oil executives, lawyers, and accountants.

Ursula and I rented a car and travelled to the Hamburg area and (West) Berlin to visit with relatives. Again, I was impressed with how the cities had been rebuilt. Hamburg had become of one of the greenest cities in Germany. I remember how Hamburg looked after thousands of trees had been cut down for firewood during the first year after World War II.

We returned to Munich for more business meetings. While there, we took advantage of a flight to the Canary Islands on short notice.

Where Have All the Pines Gone? Canary Islands

We landed in Las Palmas and drove to a hotel in Bajamar. I inquired about the bare hills we had travelled through. Photos had shown extensive forests of Canary pines. The area we drove through had been clear-cut in the thirteenth and fourteenth centuries by Portuguese shipbuilders to meet the demands for trade with Asian countries as well as the slave trade. We took home a Canary pine seedling. It did not survive in our climate.

The hills above the hotel interested us; we could see settlements on plateaus and some cave-like housing. We learned that Natives had lived there for centuries. Curious to check this out, we hiked up the hills through some banana plantations and scattered cabin-type homes with goats grazing around people engaged in harvesting fruit and grass. The sun was very hot by then. Further up, we got closer to a plateau with timber-supported caves. A mother with two babies sprawling on blankets on the ground waved at us. "Amigos!" We joined them and were invited to sit down on some hay bales, goats all around us. The lady offered wine and cheese. With some knowledge of Latin, we were able to communicate with her, explaining we were Canadian tourists visiting Germany. She pointed at a young man cutting grass: "This is not my husband; he is my friend." I asked about water. She pointed to the top of the hills and some areas storing water for discharge to different plateaus.

Our flight to the Canary Islands included a flight to Dakar in Senegal, Africa. The flight over Morocco and the Western Sahara along the African coast to Dakar showed a slice of the Sahara bordered by a seemingly endless Atlantic Ocean, a world of great contrasts. We landed in Dakar, the first visit to Africa for Ursula since having left Deutsch-Suedwest Afrika (now Namibia), where the twins were born. To follow customs, Ursula had to kiss the African soil upon revisiting Africa. The hot pavement at the Dakar airport made this very uninviting.

A bus picked us up; the driver spoke German to the mainly German tourists. He had studied in Heidelberg and gave us an excellent tour of the city and also part of the interior of the country.

The French architecture, dating back to colonial times, was quite apparent. Senegal became independent in 1971. During the tour of the city, we were told the native Senegalese are the "blackest in all of Africa." The luscious area around the government buildings impressed me. The downtown and suburban areas showed a lot of poverty. I tried to take a photo, but a tall black man put his hands on my camera. "You take these photos back to Europe; not fair!" He spoke partly French and German. I apologized and felt ashamed. I understood. Showing these photos in Germany or Canada could wrongly interpret their way of life.

The trip into the interior gave us a good impression of life in rural areas. Scattered bamboo-covered huts with people and children gathered around open fires, some shade trees, mostly acacia types, led the driver to explain, "The commonly owned fields of mainly corn are surrounding the village." Back in Dakar, we were served a refreshing mango drink along with coconuts and bananas. The restaurant was located in a very well-treed area of the city.

A visit to the island of Goa was quite depressing. The island was surrounded by a great number of sharks. Here, in the fourteenth and fifteenth centuries, the Portuguese brought slaves inland from different parts of Africa to be screened under inhuman conditions for shipment mainly to the United States and Caribbean Islands.

In Las Palmas, we made the connection back to Munich. It had been an education, but we left with mixed feelings.

Stay in Your Taxis—Panama

After another trip to Germany in 1976, we booked a cruise from Fort Lauderdale to Panama via Nassau to Haiti. A visit to Port au Prince was cancelled due to an uprising. Haiti offered a tour of distilleries; very cheap rum was a bad experience.

The next stop was Cartagena, Colombia. The naturally protected harbour had saved the Spanish Armada in the naval war between Britain and Spain in the fifteenth century. The British were unable to conquer the harbour. A city tour showed the natural beauty and the protective layout—like a horseshoe of the harbour—and the influence of the Spanish presence for centuries. A trip to the interior was also to be included but was cancelled due to banditos in the area.

From Cristobal, Panama, on the Caribbean Sea we had a connection by train via the Panama Canal to Panama City on the Pacific Ocean. The guide explained the history of the canal and the hardship suffered by mostly Chinese labourers brought in from China for the construction. Malaria and crocodiles cost many lives. In Panama City we were warned not to leave our taxis, as robbery was a daily experience. I was shocked that the tall wooden apartments were separated by only a few metres, with laundry hanging in between the buildings. The sight of the Pacific Ocean was exciting.

Lloyd and Linda visited Panama about twenty-five years later. Panama City had become a modern city. The Panama Canal was, by then, a huge construction project. Recently, I heard that China was actively pursuing the idea of building another canal parallel to the Panama Canal in Nicaragua.

The return trip via bus gave us more information about the proposed Panama construction project allowing for more international trade. A very emotional tour guide begged the mostly American passengers to help in the disposal of Manuel Noriega, Panama's president.

A stopover in Jamaica was very informative. We were shown a local rum distillery, farms, and rural life. The guide confirmed my feeling of a very relaxed mood among the locals. Harry Belafonte's beautiful songs came to Ursula's mind while singing in the bus.

We made a detour from Fort Lauderdale by plane via Atlanta to Charlotte, North Carolina, to visit my brother Hans' and Inge's family. I was surprised when the three girls spoke with a strong North Carolinian accent.

A car trip to their weekend place in the Blue Ridge Mountains, a beautiful spot with a view of the Appalachian Mountains of Virginia, was a great experience. We also visited the historic city of Charleston in South Carolina.

The Oil Business in Full Swing

Back home, I had my work cut out. Our three-person office established contacts in the Calgary oil business and introduced principals from Munich—who were interested in exploration and production of oil and gas in Alberta—to junior oil companies as well as to lawyers and accountants

I arranged a lunch meeting with Mayor Rod Sykes to acquaint the investors with Calgary's history. They were given a show – and-tell of possible investments. They arrived in busloads— groups of fifty people including investors, wives, and consultants—for meetings with representatives of oil companies for trips to Alberta's oil fields and tours of scenic places, like Banff National Park. We made our place in Bragg Creek available for BBQs with steaks on open fires and Western music. The twins entertained with guitars, singing Western and folk songs. Ursula translated German songs into English for many such occasions. Some investors wished they could have stayed longer.

I participated with experts in the oil business, on a personal basis, to invest in high-risk but well-researched projects. I regretted this later on—I lost a lot of money and it was a very costly and soul-searching experience which haunted me for some time.

The business expanded and our office grew to a staff of sixty-seven employees. Interviewing people of different professions and employment backgrounds was an interesting experience. Interviewing and selecting staff for clerical and professional positions, as well as working with people, made me aware of the challenges of this process. I observed over time that some people moved ahead in their respective positions faster than others due to their ability to follow their natural, inherent drive and talents to succeed.

I opened an office in Houston, Texas, with a staff of locally trained field and office personnel for investments in U.S. oil projects. We acquainted investors with respective projects and met with joint venture partners and landowners. We also arranged field trips. The blazing Texas sun and lack of shade presented challenges for some people. Elaborate BBQs on haciendas and armadillo races were a lot of fun. Several days of sightseeing to historic San Antonio and Austin were highlights for the investors.

Ursula and I made many friends with investors from Germany and flew frequently to Munich and other places in Germany.

Enjoying the dunes in Arizona, 1998.

The twins in Yuma, Arizona, 1999.

A dispute between the Alberta provincial government and the Canadian federal government about ownership of natural resources put an end to the attraction of investing in the oil business in 1981. The assets were sold and the companies dissolved.

This development allowed for a change of pace from the oil business. Ursula's preferred place for a holiday and change is Hawaii. A group of local bridge players made several trips to these beautiful islands. We enjoyed the atmosphere on Hawaii, the rhythm of the waves, the variety of the islands.

♦

THE WAYS OF THE WEST

Building Your Own Log Home?

A little unrest followed a call from Bob Fullerton in the fall of 1976. "We want to sell that quarter on the south side of the road closer to the hamlet. Are you interested?" Herbert and I walked over that particular geography quite often. The plateau-type layout, the mixture of pines and aspen, and the mountain view and natural clearings made this a very interesting piece of land.

Bob gave us as much time as required to make the decision. A few weeks of considering the pros and cons of this offer added some anxiety but also excitement to our lives.

The land in question had no fences. We discussed moving our horses to a neighbouring ranch, but that was not feasible. Early frost and heavy snow would prevent us from building fences on the new land. It was a very hard decision—especially for Korina—to sell our horses, even though our children had moved into the hamlet by that time.

We met the conditions of purchase. Two mobile trailers served as temporary accommodation until we completed our new homes. Rob Teghtmeyer of Bragg Creek had no problem in the middle of winter providing the necessary infrastructure, like electricity, natural gas, and water lines to the site—an enormous challenge! Rob's wife, Barbara, formerly Elsdon, her mother Mary, and husband Jack Elsdon were our first acquaintances in Bragg Creek and had contributed greatly to making us feel comfortable in the community.

Ursula and I became motivated to build a log home with pine logs from this acreage. We started out with some logs to peel and realized the amount of knowledge and time required to build with logs. We contacted Alan Mackie of the Mackie Log Building School in Prince George, BC. A trip to Prince George in November 1976 was in itself an adventure. A blinding snowstorm near the Columbia Icefields on Highway 93 turned into a whiteout; travel almost came to a standstill.

Alan Mackie left and Siegfried.

When we arrived in Prince George, we were given an overview of how to build a log home. Alan suggested a log-building course on our acreage for one week in June 1977, the first ever held in Alberta. I had to get the logs to the site and twenty students, with their own tools, to commit to the course. Alan would not guarantee the work by the students. He recommended building a twenty-by-thirty-foot picnic shelter as a project.

We were ready when Alan arrived. Some students had brought their tents along. Herbert and I made a deal with Don Wolcott, our neighbour, to acquire a three-ton truck with a lifting device. It was a hard bargain with Don over a bottle of Scotch.

The course went well. After one week, the log walls were completed to ceiling level with the roof structure on the side. This log structure served as our future living room. Wayne Sparshu and another student assisted throughout the summer to move the building onto a basement foundation for our new building site. We also built a storage shed for building materials. It was improved later to accommodate large picnics for neighbours, friends, and investor groups, capable of sheltering more than fifty people.

Lloyd took a challenging position as a fire-lookout on Falls Mountain, northwest of Calgary, for the summer months and could not attend the course. Ursula and I visited him during a long weekend. This involved a walk through the Ram River. We hung up our hip waders on a tree only to find them chewed up by squirrels upon our return. The following year, Lloyd took another assignment on Blackstone Mountain. He took his two-month-old Siberian Husky along, trained her, and named her Kiska after the Native word for mountain sheep. In the meantime, Lloyd took a great interest in log building and later taught courses with the Mackie's log-building school and became a well-known international log builder. Our home, the additional structure to the living room, was completed by Lloyd's Moose Mountain Log Homes Inc., in 1981.

Holding onto the Horse's Tail

In 1980, I took part in a twelve-day trail ride from Bragg Creek to Coleman in the Crowsnest Pass, near the U.S border, via Elkford, BC. There were five participants: Horst Hackforth, our pathfinder and organizer; Heinz Mueller, a photographer from Munich; Ken Cummings, an experienced outfitter with MacKenzie, from Banff; Bruce Macdonald, a rancher and experienced horseman from Nanton who provided some oats and supplies at various spots along the route; and, last but not least, myself, a "Saturday afternoon cowboy." The diamond hitch, the skill of tying up a pack on the back of the horse, was quite a learning experience! We were told we were responsible for seeing to it that the pack be tied down properly; otherwise it could tumble down in steep terrain.

Sig Beckedorf   Horst Hackforth

Taking a break on the trail, Sig Beckedorf and Horst Hackforth, 1981.

Checking the pack.

Leading the horse on steep terrain.

Meeting of the minds.

Saddle-weary

After a 12-day, 220-mile from Bragg Creek, west of Calgary, to Bohomolec ranch west of Coleman, five men posefor picture before begging ride to nearest tavern.Eight horses made trip. From left, Bruce MacDonald, Nanton rancher; packer Ken Cunningham, Bragg Creek; Sig Beckedorf, Bragg Creek; Heinz O. Mueller, Munich, Germany;Horst Hackforth, Calgary. Latter three are in advertising or related fields.

The "Most Wanted" at the end of ride.

A trial run through the fairly wild Elbow River after some heavy rains was a sad experience. Bruce was riding in front of me in the river when his horse stumbled on rocks and leaned over. Bruce managed to get off the horse and out of the stirrups and to jump to shore. Apparently, the horse had gotten water in his ears and lost his balance. We followed the horse downstream on horseback on both sides of the river for many miles until it came to a stop on a gravel bank with the saddle still on. A sad sight!

It had drowned. Bruce had been looking forward to training the young horse on this trip.

The first three days we had to put up with steady rain. Our rain gear served a good purpose. We set up our lean-to tent, built a fire, and managed to stay dry. With a shot of rum, diluted by rain washing off our tent, we enjoyed our first dinner. It was not until the third camp that the weather changed to brilliant sunshine. The change to sunshine from steady rains reminded me of our camping trip through Washington State with our two families on the way to San Francisco. By now each of us followed our roles in setting up and disassembling camp.

On our fifth day, we camped at the source of the Oldman River, known for the presence of grizzly bears in great numbers. This meant we had to cobble the feet of the horses to prevent them from running away. Ken told us that horses and grizzly bears don't get along. We took turns every two hours during the night checking on the horses. The next morning, Ken went into the shallow pond nearby to catch rainbow trout. He did this by using a twinned willow stick and spearing the trout. He put them in his pocket and came walking back with a smile, seeing us watching his every step. He asked us to gather some wild onions from all around the pond for a tasty dinner. While picking onions, I pointed to the glitter on the onions and grass and said to the guys, "Isn't this beautiful? So many diamonds around." I think my friends were too hungry to think about my observation.

Heinz was busy taking photos and was surprised to always have enough water for the horses. "This country is unbelievable in its beauty and resources." Many times during the ride, Ken stopped the caravan, got off his horse, and threw a rock into the brush along the way. He was lucky and skilled enough most of the time to kill a grouse, a nice change in our menu!

On our tenth day, we rode at the highest elevation, where we crossed into British Columbia. When we came to a steep hill, Ken told us, "You have to get off your horse and hang on to his tail.

The horse will pull you up. Don't worry; the horse is too busy to bother you on the way up." It worked as he'd suggested.

The next day we were about fourteen hours in the saddle trying to reach a spot near Elkford to camp. It was dark when we got to the clearing in the forest. We set up camp, built a fire, and had a good meal. Almost stumbling, we fell into our sleeping bags, mine on some rocks, which I had to remove during the night.

Early the next morning, we woke up as a forest warden shouted at us, "Are you crazy? You are only a short distance from a dynamite shed. There is road construction going on here! Get the hell out of here!"

It took us only a few minutes to dismantle the camp, mount our horses, and rush back on the trail. We called this camp, "Dynamite Camp."

After two more days straight south toward the U.S. border, we reached our destination. We had arranged to stay at a ranch near Coleman. The following day, Don Sanders, a neighbour and rancher from Bragg Creek, picked us up. The rancher in Coleman greeted us: "You guys wait a minute; I want to check the feet of the horses first!" He praised us and invited us for some free beer and a steak dinner.

The next morning while we were getting ready for Don's trailer to load up the horses, a local newspaper guy took a photo of us with the horses. "Saddle weary" with five bearded guys looking like "most wanted" subjects with our names in the paper was a nice surprise.

Forest Park

Ursula named the twenty acres Forest Park building sites.

A joint venture between Moose Mountain Log Homes Inc. (Lloyd and Linda) and Becco Resources Ltd. (our consulting company) developed a parcel of land in our subdivision. Becco looked after the subdivision and infrastructure and built access roads and drilled wells. Moose Mountain built log homes on these parcels.

In the fall of 1985, Ursula and I took on the planning and design of the land with the help of an architect from Frankfurt. The first home, a large six-corner log house, was designed and owned by this architect, a client of Moose Mountain Log Homes.

The first challenge was clearing the land for access while retaining privacy for each lot. Ursula and I decided to take the time to handle this challenge during the coming winter, tree by tree, in order to maintain the beauty of this superbly treed land.

The children presented me with a nice present for my birthday the same year: one 1964 International Harvester "corn binder" half-ton truck! This amazing vehicle, named Bruno, came in handy for handling the fallen trees, moving snow, and creating piles of wood for burning. A few dents to the truck and damaged doors were sad experiences. The main access to the subdivision had to cross some very low muskeg-type land. Jim Fullerton filled the access road with ten truckloads of pit-run gravel.

Forest Park is now an address for log homes with privacy surrounded by an evergreen forest.

Moose Mountain Meadows

Korina suggested to Linda and Lloyd that a meadow full of wildflowers with a panoramic view of Moose Mountain would be the ideal place for their wedding. It was a wonderful scene: Moose Mountain in its majesty in the background, the twins playing their guitars, and an Alphorn filling the valley with beautiful sounds! It was an unforgettable experience for families and friends spread out on the meadow!

Linda and Lloyd's wedding ceremony on Moose Mountain Meadows, 1988.

Jean-Guy, Linda's father, and I prepared the site for this occasion the evening before. While setting up outdoor toilet facilities, thunder filled the air and strong rain poured down on us. We consoled ourselves—the weather may change suddenly. Indeed, bright sunshine greeted us the next morning.

Blue Skies and Sunshine

Korina was successful in the real estate business. She purchased a property with a cute, well-kept log cabin built in 1929 by Tom Fullerton, a brother of Stanley Fullerton. The one-acre, treed lot was a good investment with residential and commercial development potential. Her contributions to preventing the construction of campgrounds on the outskirts of Bragg Creek and keeping the hamlet green were substantial.

Korina's love for things Spanish inspired her to add some Mexican decor to the cabin. We enjoyed many BBQs around an open fire and delicious Spanish cooking there. Its location on the edge of the commercial centre of the hamlet offered a good spot to watch the annual Bragg Creek parade.

Korina moved to California with Bill McMahon. Ursula and I visited them in San Clemente. What a wonderful place! We toured Capistrano Beach and Hollywood.

Three years later, Korina and Bill planned their wedding at our place in Bragg Creek. The wedding was a wonderful experience for the many McMahons and Beckedorfs.

Korina's cabin, 1996.

Handing over the bride.

Bill and Korina's wedding at our Beck 'N' Busch No. 4, 1996.

The parents.

Blue skies, bright sunshine and a background of tall blue spruce trees created a beautiful setting for a colourful crowd! Walking down the aisle (our wooden sidewalk in front of the log home) with Korina to "hand her over to Bill" was an emotional experience. Ursula and I have great memories of this day.

♦

NEW CONTACTS

The contacts made in international business offered opportunities to work with investors. Frank Stempfer, a computer expert and businessman, emigrated from Austria to Canada. I met him in Calgary. He invited me to join him to explore international business opportunities. In our spare time, Frank and I enjoyed selling small blue spruce trees from our nursery off our IHC "corn binder" truck in areas around Bragg Creek and Springbank. Frank demonstrated a sense of good humour, an excellent work ethic, and good salesmanship.

Some old contacts, private and business-oriented, remain active to this day. A friend in Hamburg is prodding me to eventually translate my "book in progress" into German with his help.

Mining in Mexico

Frank Stempfer and I met with a geologist in Tuscon, Arizona, and flew to Hermosillo in the state of Sonora, Mexico. We travelled east by pickup truck into the Sierra Madre Mountains, close to the border with the state of Chihuahua. A very rocky road took us into a mixture of jungle and sparsely covered hills. We were warned of banditos along the way—but experienced no such adventure. A German lady introduced herself as the owner of the mine in partnership with her husband, a Mexican businessman who lived in Mexico City. After a real Mexican treat that included a great variety of burritos, a business plan and maps were spread out. Our geologist asked for drill samples and production data. "We have that at the mill," we were told.

The mill was operating with material from the surrounding surface areas; no drill samples, only gold and copper nuggets. Records showing sales to a firm in Mexico City were produced the next day.

Our accommodation was a large shed with bunk beds for employees and an extra room for visitors. I was very tired and was awoken early in the morning by the seemingly endless crowing of roosters. A spicy Mexican breakfast with a view into the surrounding pine-like forest beyond the clearing was a nice surprise. Looking at the dew on the leaves and grass surrounding a waterhole reminded me of diamonds everywhere. Further inspection of records available and the amount requested for participation, however, did not meet our expectations.

We visited more opportunities in the mining business in California and BC for presentation to German business interests. I appreciated Frank's persistence and entrepreneurship, and we maintain a lively friendship and business partnership to this day.

But Where is the Water?

A German businessman contacted me with a business plan for a potable water plant. He suggested the Canmore area for prospective sources of water. But the proximity of coal mining shafts in various locations of interest discouraged us from pursuing this opportunity.

A visit to the Nanton Water plant was very interesting as this was a going concern and a good success story. An investment was offered for the extension of a pipeline to an area west in the direction of nearby Porcupine Hills with a proven extensive water supply. A feasibility study was in progress. A meeting with the engineering firm disclosed that the distance for the pipeline was much greater than anticipated though, making the project unprofitable.

At this time, I researched the subject of water, our most precious resource. My interest continued while travelling in foreign countries. In Germany, I found excellent worldwide research about environmental effects on the health of all living matter as well as the misuse of water resources by industry.

Greenhouses and Gas Plants

I was introduced to a Dutch couple, the de Jong's. A successful operation in the Netherlands inspired the De Jong family to look at flared natural gas in Alberta gas plants for heating and cooling greenhouses at reasonable costs.

I made contact with an engineer at Shell Canada, but learned that Shell's plants were in locations not feasible for greenhouse operations. He referred me to Imperial Oil and suggested their plant at Quirk Creek near Millarville as a good location.

Imperial Oil was very interested; a visit to the plant encouraged us to prepare a business plan for a glass greenhouse, advantageous for the sunny exposure in that location. We received information on access to their natural-gas dispensing facilities and costs for modifications.

It was interesting to work with the de Jongs. Arthur de Jong prepared a plan for growing lettuce and tomatoes; I followed up with marketing research within a hundred-kilometre radius of the plant. The availability of water was a challenge; neighbouring ranchers had some concern.

When we had gathered all the necessary information, we determined that the costs of access to natural gas for the greenhouse structures exceeded the expectations on both sides. Other locations did not seem economically feasible, either. Eventually, I was able to assist the de Jongs in settling in Vancouver to set up operations.

"Success is not a future event.
Success is your state of consciousness . . .
to be totally alive in this moment."

—Eckhart Tolle, *A New Earth*

♦

TIME TO TRAVEL

The words "land in sight" on board the *Karl Grammersdorf* in 1951, when I looked at the shores of North America for the first time, were overwhelming for me. It was an awakening of a natural instinct to look for greener pastures; for our ancestors, maybe a drive for survival. Ursula and I shared a great interest in exploring living conditions and lifestyles in other countries.

We were looking forward once again to combining business and travel as an opportunity to see the world. The Calgary Chamber of Commerce luncheons brought people together with interests in various fields of business in foreign countries.

Venezuela, Germany, Russia, Poland, and Hungary

Two trips to Venezuela gave Ursula an insight into life in Caracas and Barquisimeto. The parents of a friend in Calgary had moved from Germany to Venezuela before World War II. The war held them there. Beatrice Meyer and her husband, Rudy, moved to Calgary in connection with the oil business. Ursula was invited to join Beatrice when she visited her parents and friends in Venezuela. Ursula spent a lot of time with Beatrice's mother, walking, singing, and accommodating her. She suffered from Alzheimer's disease.

Excursions into rural tropical areas and the luscious island of Margarita demonstrated a contrast of poverty and wealth.

I had the opportunity to visit Germany during this time. At a luncheon at the Calgary Chamber of Commerce, I was introduced to the principals of Alternative Fuels Systems Inc. (AFS). Their dual fuel technology allowed natural gas to be injected into diesel engines in trucks to reduce diesel fumes, especially in large cities.

AFS engaged our consulting firm, Becco Resources Ltd., to explore the marketing possibilities in Germany. This was the beginning of frequent trips with AFS engineers and principals to Europe. I prepared visits to offices of major German manufacturers of trucks, cars, and automotive products, in addition to large cities in Germany. For this purpose I contracted Dr. Ferdinand Herms, a diesel engineer I had met at a conference in Berlin. Dr. Herms was a recognized specialist in this field with a large staff of engineers working with the German auto industry. After World War II, he had instructed Russian engineers in diesel technology in East Germany.

Dr. Herms opened many doors for us for presentations of AFS technology. Ford Germany, Leverkusen, Daimler Benz, Stuttgart, and MAN (Munich, Augsburg, and Nuremberg) as well as municipalities in Berlin, Dresden, Chemnitz, and Hamburg expressed strong interest.

Extensive research led to frequent presentations and repeated visits. I worked out of an office at the residence of Ursula's cousin in Berlin whenever I had to stay for a longer period.

I observed an interesting interaction between the German and Canadian engineers and the businessmen at our conferences. The Germans introduced a process requiring the firm we visit follow a strict agenda. For instance, Dr. Herms introduced the Canadian participants first and their technology after, only in summary since a written outline had been presented earlier. E.g., the executive of MAN introduced his firm followed by a history of the company with reference to Rudolf Diesel, the inventor of the diesel engine in the nineteenth century. Rudolf Diesel was part of the early years of MAN. The Germans' expectations of a potential joint venture or partnership were laid out in great detail. Frequently, Canadian representatives were somewhat impatient with the lengthy process of the meetings and having no chance to follow their own agenda. Dr. Herms made the Canadians aware that this was necessary procedure in order to become more acquainted with a large and worldwide potential business partner. My experience with German investors in the oil industry in Canada in the 1970s is that they followed a similar agenda with their potential Canadian business partners. The success of the meetings AFS held in Germany depended very much on understanding the agenda beforehand. For this purpose, I emphasized the importance of having Dr. Herms at these meetings. The lack of success in Germany to some extent was based on AFS technology's being ahead of its time as well as AFS misunderstanding the necessary agendas to follow.

My contacts in Germany extended my visits with German investors and Canadian engineers to Russia, Hungary, and Poland. In these countries, a lack of willingness or ability of the foreign firms to participate in sharing the funding was the main reason for missing the mark.

A visit to Russia followed a meeting I had arranged earlier with a German firm active in Russia's building of natural-gas service stations. Lurgi Anlagenbau of Chemnitz in former East Germany agreed to join us in presentations in Moscow and Yaroslavl.

When I arrived in Moscow, I checked into the Cosmos Hotel, where I met with Gerry Klopp, the chief executive of AFS. The lobby was filled with guests mainly from Kazakhstan according to our interpreter. At the counter I was told to hand over my passport. I hesitated, but the interpreter informed me that this was a necessary procedure for security purposes; the passport would be returned in the morning. I felt OK since I carried a copy of my passport data when travelling to foreign countries. The Cosmos Hotel was the entry point for most citizens of former Soviet Republics.

At Red Square in Moscow, Russia, February 1993.

Red Square.

Red Square in morning light.

I found our meetings in Russia interesting but slow. Our group usually consisted of about six people including an interpreter. The Moscow meetings involved thirty Russians: businessmen, engineers, and regional government representatives. None of them seemed authorized to make decisions. A lot of time was wasted waiting for results of our presentations. We used the time to tour Moscow and visit the Kremlin and rode the amazing subway Stalin had built. The walls in the hallways were covered with marble tiles.

I often enjoyed invitations from our prospective business parties in foreign countries after meetings on a private basis. The Russians were very reserved in this respect. A former Russian military officer in his role as our interpreter tried hard to overcome the attitude of the Russians. It was important to greet our foreign partners with a lot of respect for their culture.

An Ikarus bus took us to Jaroslavl, a "forbidden city" during Stalin's time due to the extensive diesel manufacturing plants for the army there. A seven-hour trip northeast of Moscow included a stop at a service station in Rostov. The bus could not get out of the parking area due to sheer ice. A tall farmer in furry clothing arrived in a tractor and pulled us up to the road. I mentioned to our interpreter that I had not expected to see "the famous Russian bear." He replied, laughingly, "This is the only entertainment we can offer." It was February, the coldest time. The rural areas northeast of Moscow along the way looked very bleak, in deep snow. Scattered villages did not show much life; windows were shuttered to save heat. We were told the residents maintained root cellars to survive the long winters.

Finally, in darkness, we arrived at our destination—a log hotel. It was started in the tenth century and built over a long period of time by Norwegian craftsmen employed by the Czars. The logs were placed on boulders to keep them dry and painted over many times with reddish brown paint. This allowed the logs to serve their function for centuries.

Each one of us had an assigned suite. Creaky wooden steps took us to the third floor. I could not believe my eyes when I saw curtains flying in my suite! Apparently, windows had been left open by mistake and now the water pipes were frozen. The staff member shouted. I couldn't understand but followed him downstairs. It took quite a while for someone to appear to fix the problem.

A man sitting in the lobby waved at me and asked in German, "Would you like to join me for a drink?" I shook his hand and thanked him. He sympathized with my disappointment. I heard an interesting introduction: "I was born in Russia of German parents. I am selling German diesel parts. How about a shot of vodka?" He had overheard the excitement about frozen pipes and our planned visit to the diesel works in Jaroslavl. "You have to be very patient with the staff there. A slow-moving outfit. How do you fit into this? From what I hear, this is a Canadian group that checked in?" I explained my involvement. Finally, I heard that I could get into my room.

The next morning, we had a good breakfast of Russian dark rye bread and eggs and bacon; even a shot of vodka to add spice. Then we visited the diesel plant. On the way, we passed a large log structure in a park-like area surrounded by tall spruce trees. The onion-like towers looked mysterious. This majestic structure was explained as being the first Kremlin built in Russia, about the same time as our hotel. It looked well maintained. The Mongols destroyed the Kremlin in Moscow during their invasions over many centuries. Jaroslavl had been missed due to its location.

The diesel plants were quite extensive but the buildings looked neglected. Some forty thousand people were employed here during the war. There were fewer than ten thousand at this time. We were served vodka and sausage and very bad coffee. The meetings, like in Moscow, were not well organized and took many hours with not much feedback from the Russian group. We left fairly disappointed as we were told a decision would be made in Moscow. Walking through the seemingly deserted streets of Jaroslavl on a clear and very cold day, diamonds on the snow-covered trees made me feel OK. A new mink fur hat felt comfortable. Heavily dressed residents moved slowly, not paying much attention to the visitors.

We returned to Moscow and used this time for sightseeing. A taxi took us on tree-lined streets. I was surprised to see hundreds, maybe thousands, of crows sitting on bare branches of elm, birch, and black oak trees. We looked at Russian-built, former military trucks converted to heavy vehicles for the construction industry. Most of them were run by diesel engines manufactured in Yaroslavl with parts from Germany.

The taxi driver who took me to the airport in Moscow for my return trip to Frankfurt spoke German. We arrived at the place where the Russians had put large steel barricades on the way to the airport to hold back German tanks as they approached Moscow. The driver pointed out, "That is as far as the Germans came when they tried to invade Moscow. Stalin had already left the city. Our cold winter did the Germans in."

It was not until we got back to Calgary that we found out the Russians' proposals were not acceptable. A counterproposal was considered.

An assignment awaited me in Poznan, formerly Posen, in Poland. Insurance coverage was not available for leased cars when travelling to Eastern countries because of vandalism, so my friend in Berlin offered to drive me. The City of Poznan was planning the installation of the AFS System in a VW van. The negotiations were successful. No interpreter was needed because the technical people spoke German.

On the way back to Berlin we took a scenic tour via Danzig and Stettin, now part of Poland. There was a lot of renovation activity, as both cities had suffered heavily during the war. This part of the country was the home of my ancestors dating back to the eleventh century. I was surprised how well German was spoken in restaurants and hotels.

I spent some time with relatives in Frankfurt while preparations were made for a meeting with officials of Dresden. The park-like setting of a bed-and-breakfast place inspired me to bring my journal up to date.

Sometimes, after a long day of meetings and travelling, I felt tired and restless. With "checking in" (as I call meditation), I can let go. Totally relaxed, I direct my inner energy from head to toe. I let it flow, feeling good.

Gerry Klopp of AFS joined me in Frankfurt. We took the train via Passau, Vienna, to Budapest, a scenic country along the Blue Danube. AFS worked for some time with a diesel engine firm in Hungary, providing engines for the Ikarus bus. This vehicle was very popular in Eastern Europe, as I experienced in Russia.

I looked forward to learning more about Hungary's origins. The language is not Slavic, even though Hungary's origins can be traced back to the eastern part of the Ural Mountains.

Most cities of Europe I travelled to showed traces of World War II. Budapest surprised me; there were no traces of recent wars. Medieval architecture showed a rich artistic and historic heritage and natural characteristics. The city is divided by the Blue Danube.

Government officials met us in Budapest for lunch and discussions of business in Hungary. Each one of us was presented with a colourful book, *Hungary*, which explained the history of the city and the country. The origin of the Magyars can be traced back to at least 2000 BC. Nomadic hunters in the area of the Volga River and the Ural Mountains migrated west and, around 896, settled and founded the principality of Hungary. Later, the Kingdom of Hungary became a Christian bastion against the Mongols and, later still, against the Russians. In the seventeenth century, the northwestern part of Hungary came under Habsburg rule, while the south fell into the hands of the Ottoman Empire.

In the evening, we were introduced to a restaurant featuring menus of the Donau Schwaben, who emigrated from Swabia in southwest Germany in the seventeenth century, following the Danube (Donau) to Budapest. I was sitting next to a Hungarian government official. I mentioned that I was of the opinion that Hungarians and Finns were related and had their roots in Eastern Siberia. He confirmed that, at one time, historians thought this to be the case. The most recent evidence shows that the Finns appeared after the last ice age east of the Baltic Sea and dominated the country of today's Lapland and Eastern Sweden until they settled in their present location. Their origin is mostly of Scandinavian descent.

China

Dr. Ralph Ashmead, a Calgary agriculturist, made contacts in China for improving milk production there by using
Canadian know-how.

Ralph asked me for assistance in finding partners in Germany
for the production of yogurt in China. Some time before, I had
met a German engineer, Hubert Kulmus, who was employed by
Suedmilch AG, in Stuttgart, one of Germany's largest producers
of dairy products, including yogurt. I made contact with Hubert
in Stuttgart. He had opened his own engineering firm, consulting
to Suedmilch. I visited him in Germany and made preparations
for a meeting in Changsha, Hunan Province, China.

I was very excited about my first trip to China, "the land of
eternal smiles and mysteries." After a night in Hong Kong (still a
British colony at that time), we had to pass through the custom
zone for a train ride to Guangzhou, Mainland China. Walking
through a great number of turnstiles, I was overwhelmed by the
huge crowd. I lost sight of Ralph in the crowd; he was held up at
the turnstiles and showed up at the taxi stand. He had almost
lost his laptop.

At the airport, I observed four young, tall Chinese men dressed
in business suits. They marched straight to the first-class line-up.
After a two-hour flight from Guangzhou to Changsha, we met
Hubert Kulmus in the hotel. Here, we were greeted by a regional
government delegation including Chinese I had observed in
Guangzhou. We learned that the Chinese government owned
the hotel and that the four well-dressed Chinese represented the
military. It felt strange to parade through the large lobby of the
hotel in this military-like reception. Hotel guests looked at us in
surprise. An oversized meeting room was available with about
thirty Chinese government representatives and business and
military people around a long table, with flowers lined up down
the middle. I was in awe. Each of our Chinese participants looked
at us with intense interest. This region of China did not as yet

have much contact with business people from the West. The head of this group made a good effort, through an interpreter, to introduce each of us, including our background and position, to the group. An agenda for our three-day stay was laid out. I felt comfortable with this group.

Meeting in Shanghai, China, 1993.

Early the next morning, I went for a walk around the hotel. I had the opportunity to join a tai chi group in a parking lot; they hardly noticed me during their silent exercise. After breakfast, a large Mercedes limousine took us to a rural area, four hours northwest of Changsha. It felt like a mafia car. This was a government vehicle used to impress us, I think. Initially, we travelled through lower regions where farmers with large straw hats followed water buffalo pulling ploughs. The driver told us, "Like it was a thousand years ago, they own small rice fields." We passed through higher regions with tea plantations until we reached another low-lying area dotted with small dairy farms.

In order to reach the fragile-looking farm houses surrounded by partly flooded fields, we had to walk on wooden walkways. A mother greeted us with a baby on her arm. She smiled and did not say anything. The baby stared at me with big black eyes. I heard the interpreter saying, "They have not seen any people from the West and your white beard is quite a curiosity for the baby."

Rural area north of Shanghai, 1993.

Visiting dairy farms in Western China, 1993.

An inspection of the nearby dairy herd was quite disappointing to Ralph: "What are you feeding these cows? They look very thin." The answer was grass and hay, sometimes supplements. A milking demonstration revealed poor milk production. Tests of the quality were considered acceptable but the barns did not meet Canadian standards. Hubert showed a lot of concern about the health of the dairy cows.

Ralph and Hubert completed their report with details of the disappointing results. Ralph suggested having Alberta professionals visit this area. Alberta was well known for imports of quality dairy livestock semen. The regional government agreed and put us in touch with its counterparts in Suzhou. A meeting was proposed for the following year in Shanghai and Suzhou, "the Garden City of China." We were encouraged, as apparently the area was well known for its extensive agricultural activities. It was mentioned that Danone, the French yogurt-manufacturing firm, had visited Suzhou a year ago.

Before we left China, we met with two representatives from Suzhou to prepare for our visit the following year. Hubert advanced the idea of second-hand, guaranteed, stainless steel equipment for a yogurt manufacturing facility. The Chinese were quite excited, knowing quite well new equipment would not be considered for economic reasons. Hubert had a great sense of humour, and won the confidence of the Chinese in a very short time. He was well prepared for this visit: he laid out the plans and guaranteed used dairy equipment and installation procedures.

The next year, I landed in Shanghai and met with Hubert and Ralph. We toured the city with an interpreter. In 1996, Shanghai was "one construction site." I wondered about huge scaffolds made of bamboo. "These structures will gradually be replaced by steel, but we don't have enough of that at this time." One skyscraper had a huge Volkswagen sign installed at the very top. "Volkswagen employs and trains a lot of Chinese workers; China supplies the materials for the large plant in Shanghai and the Germans provide the know-how," our interpreter explained. All taxis in Shanghai were made in Spain (seat model) by Volkswagen. I had another opportunity for an early morning tai chi exercise on a walk around the hotel.

On the way to Suzhou (in the direction of Nanjing), we travelled through many cities I could not find on my map. I was told that only places with over one million people were shown. We stopped for lunch in a country tavern and walked along a pond stocked with fish. The water did not look inviting to me. In a separate open room inside the restaurant, some Chinese people were singing karaoke. What a pleasure! We enjoyed the deep-tone singing many times on our travels through China.

Suzhou, one of the oldest cities in China, has a population of about two million people. The surrounding fields of vegetables looked well maintained. Harvesting was still in progress. This "Garden City" earned its name. Tree-lined streets and a lot of flowering shrubs greeted us as we drove to our hotel in a suburban area. What a difference from the places we had visited the year before.

An assembly of regional government and business people welcomed us in a large convention room. Two people we met the year before introduced us. It was a very friendly and open atmosphere.

Shanghai meeting with regional government and businessmen, 1996.

The hotel displayed a strong Chinese decor. Accommodation was excellent; always service with a smile. We got right down to business after breakfast with our Chinese partners, again with long rows of beautiful flowers lined up in the centre of long tables. A van took us on a tour through the city and into rural areas. Large, government-controlled operations showed good management, along with what looked like family-run farms.

First, dairy herds were inspected and milk samples taken. A visit to stores

that sold dairy products disclosed some yogurt products by Danone. Hubert picked up samples.

Ralph and Hubert laid out a prospective yogurt-production facility in meetings that followed. We were surprised how quickly our Chinese friends reacted with sincere interest and detailed questions. On our final day, we signed a memorandum of understanding. Hubert assured a quick response from Germany, subject to approval by the federal government in Beijing. A complete list of available equipment, costs including installation, as well as timing of shipment from Bremen, Germany, to Shanghai and on to Suzhou would be forthcoming. An approval by Beijing was expected in short order.

A beautifully decorated dining room and a Chinese buffet with a huge selection made us feel optimistic about our future joint-venture expectations. The seating arrangement allowed me to sit next to a Chinese engineer. He spoke English fairly well. I had talked to him before when he told me his daughter took English in high school and had taught him. He liked our business proposal and was confident that the federal government would approve it. "Canadians seem to be generally appreciated for their approach to business. They appreciate our culture and are polite, not arrogant." He told me that three generations lived in his home, quite common in China. I asked him about entrepreneurship. I heard previously that a growing middle class in China could mean more independence in business in general. "Not for some time," he replied.

The head of the Chinese group offered a toast to success in our business with a local rice schnapps. We said goodbye to each one of the twenty-eight participants at the dinner.

Four months later, we were informed that the Chinese federal government had decided to ban the import of used food equipment and machinery into China. The regional government in Suzhou appealed this decision several times, without success.

Shanghai, China, 1995.

Western China, close to tea plantation near Suzhou, 1995.

Suzhou in Hunan Province, 1995

Meeting in Suzhou, 1995.

The flight plan from Shanghai with a stopover in Beijing across the Sea of Japan and the Pacific Ocean to Vancouver, was changed drastically. Above Russian territory, the pilot announced, "We have to lighten the plane and drop fuel due to the failure of one engine. We will return to Beijing for repairs." After one hour inside the plane at the airport with no air conditioning, we began to feel quite uncomfortable. Finally, we were told hotel reservations had been made in Beijing with all expenses paid and that we would be re-boarding in the morning. We had the opportunity for a brief tour of Beijing before our flight continued in the afternoon with a stopover in Tokyo to Vancouver: a thirty-six hour delay.

Science Under the Microscope

During my travels in Europe and China, I had opportunities to discuss various subjects during lunch, dinner, and the weekends with engineers and physicists. At times, I could not follow their reasoning, but their enthusiasm and excitement drew me into listening with a non-scientific mind. I was inspired to read Gary Zukav's *The Dance of the Wu Li Masters*. It is an overview of the "new physics" and winner of the American Book Award. It is a book written not only for scientists, but for the uninformed as well. I also like the humour throughout the book. From the introduction:

> The New Physics, as it is used in this book, means Quantum Mechanics, which began with Max Planck's theory of Quanta in 1900, and relativity, which began with Albert Einstein's special theory of Relativity in 1905. The Old Physics is the physics of Isaac Newton, which he discovered about three hundred years before. Classical Physics includes the physics of Isaac Newton and relativity, both of which are structured in this one-to-one manner. It does not, however, include Quantum Mechanics, which, as we will see, is one of the things that makes Quantum Mechanics unique.

I quote with reference to page 123:

> Heisenberg's remarkable discovery was that there are limits beyond which we cannot measure accurately, at the same time, the processes of nature are at work. These limits are not imposed by the clumsy nature of our measuring devices or the extremely small size of the entities that we attempt to measure, but rather by the very way that nature presents itself to us. In other words, there exists an ambiguity barrier beyond which we never can pass without venturing into the realm of uncertainty. For this reason, Heisenberg's discovery became known as the "Uncertainty Principle."

In a recent interview by Deutsche Welle (DWTV), Dr. Rolf Dieter Heuer, general director of CERN (LHC Collider) commented on the question of what drives him to pursue this gigantic scientific project:

> Everyone wants to know more, is curious about what holds the world together. It is an inner message, a mutual acceptance of a "grey zone or interface" between scientific fact and belief. We are at the point where we are closer to our goal. We are opening one door to find there are more doors to open—possibly to infinity.

During a return flight to Calgary, I was referred to the book *Super Brain* by Deepak Chopra, M.D., and Rudolph Tanzi, Ph.D. The subject of our brain and consciousness has occupied my mind for some time. Here are some quotes from *Super Brain*:

> Consciousness—the invisible agency of the mind—created the brain and has been using it ever since the first living organism began to sense the world. As consciousness evolved, it modified the brain for its purposes, because the brain is only the physical representation of the mind....
>
> I am an optimist, and I hope to see the validation of consciousness reach full scientific acceptance in the coming decade....
>
> Your mind is the rider; your brain is the horse.

Rudol E. Tanzi, a professor of neurology at Harvard University, director of the Genetics and Aging Research Unit at Massachusetts General Hospital, and head of the Alzheimer Genome Project, says in the epilogue:

> "The signs of progress on all fronts—chemical, genetic, behavioral, and lifestyle—are encouraging. But they alone wouldn't have led me to write about "Super Brain." In my field you can thrive by

being a superb technician, carving out your scientific niche in the detailed analysis of very narrow aspects of a disease. You can make it pretty far in science by ceasing to speculate and obeying the dictum "shut up and calculate." Hard science is proud of its status in society, but I have also witnessed firsthand that this pride can extend to arrogance when it comes to contributions of metaphysics and philosophy to developing scientific theories."

This broad dismissal of anything that cannot be measured and reduced to data strikes me as incredibly narrow-minded. How can it make sense to dismiss the mind, however invisible and elusive it may be, when science is entirely a mental project? The greatest scientific discoveries of the future often begin as pipe dreams of the past. . . .

We must never forget that the true seat of human existence is in the mind, to which the brain bows like the most devoted and intimate of servants.

Published in n2012, *Super Brain* was followed up and co-authored by Dr. Chopra and Dr. Tanzi in 2015 with *Super Genes: Unlock the Astonishing power of Your DNA for Optimum Health and Well-Being.* This book illustrates an interplay of nature and cutting-edge genetic science undreamed of a decade ago.

I consider Drs. Chopra's and Tanzi's work in their books *Super Brain* and *Super Genes* an eye-opening and pioneering effort written for everyone curious about our human potential. A teeming microscopic world and the amazing animating life within that is influenced by every thought and action is co-authored by well-known professionals in great detail.

Great writers inspired me often to look at a little booklet I picked up on my travels:

Walt Whitman:

> "Be curious, not judgmental."

Ralph Waldo Emerson:

> "To be yourself in a world that is
> trying to make you something else, is
> the greatest accomplishment."

> "Adopt the pace of nature:
> Her secret is patience."

Henry David Thoreau:

> "It's not what you look at that
> matters, it's what you see."

William James:

> "The greatest discovery of any generation is that a human being can alter
> his life by altering his attitude."

Albert Einstein:

> "Imagination is more important than
> knowledge. Knowledge is limited.
> Imagination encircles the world."

Max Planck:

> "Science cannot solve the ultimate mystery of
> nature. And that is because, in the last analysis,
> we ourselves are part of nature and, therefore,
> part of the mystery that we are trying to solve."

Eckhart Tolle, *The Power of Now*:

> "Accept the present moment fully.
> The past is gone; the future not here
> yet, let this put you at ease."

Richard Dawkins, *The Greatest Show on Earth*:

> "Evolution is within us, and its workings
> are embedded in the rocks of eons past."

Lao Tsu:

> "He who conquers others is strong. He
> who conquers himself is mighty."

Surely, a lot of food for thought!

Africa and Back to Germany

It was time to visit the twins' birthplace in Namibia, Africa. Ursula and I flew to Windhoek, the capital of Namibia, via Frankfurt. Namibia was a former German colony dating back to the 1880s during the Kaiser's time until the colony was lost to South Africa during World War I. Active tourism between Namibia and Germany continues to the present day.

Ursula was quite emotional upon landing in Windhoek. Her mother left Namibia with the twins to return to Germany in 1933. We planned to visit Tsumeb, her hometown. Our reserved car was waiting for us. "You may realize Namibia is one of the most risky places for cars due to narrow roads and shifting desert sands. Don't drive at night. Another thing, there is no insurance against sandstorms, which are quite frequent." (Avis Travel). Well, we were not aware of this. I practiced driving with the steering on the "wrong" side in the parking lot. With map in hand we got lost on the first turn when we drove into a large camp of native

Hereros gathered around open fires. Their curiosity made us turn around. Hereros are members of a tribe originating from the Tanzania area. They are tall and proud people who speak the Bantu language. We encountered them a few times on our trip. Large groups of rural families are constantly on the move to the cities looking for work. We spent the first night in Windhoek with a couple referred to us in Calgary. Herr von Stolzenberg, the owner of the apartment, maintained several hunting lodges in the area. He gave us good advice for travelling.

The next stop in a small town attracted us. A restaurant called Hamburg looked very inviting. We were served by smiling natives with white aprons. The nicely placed tables with white tablecloths were a surprise. We were asked in German to choose from the menu of local and German food. Along the way we stopped at places where wooden carvings were offered. I asked about a hiking stick with

Namibia, April to May 1995

Rented car in Windhuk; right-hand driving; caution: desert sand!

Namibia is slightly larger in area than Alberta. Its population is 1.8 million: 20 percent white, mostly of German, Dutch, and English background. Original people were Bushmen (Hottentots), Hereros, and Ovambo. The official language is English, although German and South Afrikaan (Dutch) is widespread.

Namibia was a German colony from 1884 to WWI and became an independent republic in 1990. Apartheid was not apparent in this country.

The earliest signs of human presence date back twenty-five to twenty-seven thousand years.

German missionaries built the first white settlement in 1805.

Many signs of the German Kaiser period remain in architecture, railroads, highways, the mining industry, and school systems.

German tourism is very active to Namibia with direct frequent flights by Lufthansa to Windhoek from Frankfurt.

The climate is hot during the day and cool at nights. Vegetation ranges from hardy desert plants to dense forests. There is a great variety of wildlife and the largest free-roaming animal reserve in Africa: the Etoscha-Pan.

Main industries are tourism, ranching and farming, and diamond and copper mining.

a meticulously carved native woman's head; beautiful work! I asked for the price and tried to haggle. The tall young man of the Hereros tribe replied in German, "This is my price; I want my children to travel like you." I agreed to the asking price and thanked him. *"Gute Reise* (happy trails)," he replied.

Travelling toward Tsumeb, Ursula's birthplace, we saw stout warthogs and quick pavians running and jumping alongside the road, constantly watching us with great curiosity. We now entered hills covered with small pines in contrast to the desert-like terrain we had travelled through so far. "This looks like Thueringia except for the small size of the pines," Ursula said, with reference to her home province in Germany. Signs of Tsumeb, the "railway and engineering hub," appeared. Indeed, we learned a lot about the beginning years in Namibia, then called Deutsch Suedwest Afrika. Tsumeb attracted many engineers from Germany to establish successful mining ventures and build railways and an infrastructure to serve the colony.

We stayed in a hotel known to Ursula from her mother by the name of "von Epp," friends of the family. The next day we explored the well-treed city with many German street and business names. When we visited the local museum, we asked the manager about any traces of the Buschendorfs (the twins' maiden name). What a surprise when Ilse Schatz sorted through some files and came up with a photo of a medical doctor holding up one twin in each arm with an announcement of their birth, dated January 31, 1930! Ursula found notes in her mother's journal and letters. When the twins began walking their mother had an area in the shade of the house fenced in for a play pen. She put in two of her monkeys, a piglet, and chickens for company. A photo showed that Ursula was trying to climb the fence, her hair had been cut off. Their mother had put two and two together when she noticed the twins had pulled each others' hair out and put it in their mouth, imitating the monkeys. The family moved back to Germany when the twins were three years old.

The Schatz family invited us over several times and we learned much about the early history of Tsumeb and Namibia and received valuable advice for our travels to other places. We visited the hospital where the twins were born. What an experience for Ursula!

On the way to a guest ranch, close to the border with Botswana on a sandy bush trail with little shade, we had a flat tire. We were warned of marauding Natives in this area and became somewhat anxious to get to Grootfontain, the next town, for repairs. Tired and hot, we arrived at the ranch and were refreshed when we saw that the place was like an oasis. Tall acacia-like trees embraced the flat houses of the residence, a guest house, and barns. In the background, I noticed more buildings, somewhat hidden in the trees. These buildings turned out to be the residences of Native employees on the ranch.

North of Windhoek, Namibia, 1999.

We reserved a room at there, recommended as one of the oldest cattle ranches in Namibia. The original Bayer family had settled here in the 1880s. Their Native help also lived here for generations and spoke German when we were introduced. The Bayers maintained their own cemetery.

A typical Bavarian dinner with *spätzle* was waiting for us. Other guests on the way to Botswana travelling from Germany joined us. Mrs. Bayer explained to my question about family: "Our children are attending university in Munich. They only visit us during holidays and do not want to live here." Mr. Bayer added, "Times are changing. The Natives are anxious to buy land; many settlers are selling."

The rancher took us on a day-long trip around his "big spread" in a large truck with benches. Chudop (caragana type) trees and a mixture of fruit-bearing tamboti, mopane, and camelthorn trees dotted the landscape with dry-looking meadows in between. We took a break where tall white Brahma cattle with large horns grazed in the vicinity of a water well. Bayer gave us a demonstration of filling large water containers by operating an old-fashioned pump run by a diesel generator. "This is the thirteenth year of drought. We are fortunate to have lots of water deep down here." What a smell of fresh air! I commented on the lack of wildlife. "This land is so widespread; we have lions, kudus, and gemsbok. They are spread out far. Our cattle know how to protect themselves."

We returned to a nice dinner and wine. Bayer gave us helpful tips for travel to the Etosha National Park, one of the largest open and unfenced wildlife reservations in Africa. "And don't forget to look at the HOBA meteorite on your way. It's a big one."

I stood in awe looking at the big chunk of shiny black rock from a distant star. The HOBA meteorite hit the earth about eighty thousand years ago. It was, at that time, the largest on earth, three metres long, upright up to one metre thick. Discovered in 1920, it was declared a national monument in 1955. It contains 82.4 percent iron, 16.4 percent nickel, and .76 percent cobalt, as well as sulphur, carbon chrome, copper zinc, germanium, and iridium. The scientific description reads, "Nickel-rich Ataxit." No wonder many attempts were made to scratch or cut some samples—none of them penetrated the solid mass. "I wish I could share these experiences with Omi," Ursula said as we took a photo.

Three former stations of the Etosha Pan, built by the German *Schutztruppe* (military police) in the 1900s against marauding tribes from Angola, were re-constructed to meet the growing number of tourists from many parts of the world. They were about 100 to 150 kilometres apart, requiring three days to cross the pan.

Before we entered the first location, Namutoni, we encountered giraffes strolling along the road, looking down at us from in between leaves of large trees. It seemed unreal. Namutoni offered modern accommodation, educational facilities, and excellent restaurants. The rules are to come in no later than 6 p.m. and leave any time after 6 a.m. Travelers were not allowed to leave the car and had to take photos from inside the car; game wardens were watching everywhere.

We left for the station early in the morning and after one hour we encountered a lion mauling a wildebeest with vultures circling above. Water holes were spread out along the road. Wildlife, such as zebras, congregated in large numbers any time during the day. We saw giraffes and also stout and confident warthogs. These warthogs surprised us time and again, demanding their place around waterholes. They were Ursula's favourite animal. Toward the evening, large herds of elephant, and some rhinos, lions, and zebras gathered around to take turns drinking. The scenery changed from a prairie-like setting to savannahs with small brush and camelthorn trees. We enjoyed traveling through ever-changing scenery, watching animals all the way. Ursula sang while we were traveling; this, after all, was her home country.

The second station called Halali (a German hunting call) offered nightly viewing of wildlife around a huge lit-up water hole. A high wire fence separated the audience from the animals. As we were watching three lions in a single row along the water hole, we noticed a large rhino on the same path heading for a collision. The lions and the rhino stopped, looking—we held our breath—the lions preferred to take a trail leading around the rhino! Elephants took a long time to de-water themselves, throwing water over their shoulders to cool off.

Okaukuejo, Namibia, 1999.

City park, Okaukuejo, 1999.

Grootfountain, Namibia, 1999.

Ilse Schatz, director of museum in Tsumeb, with Ursula and Siegfried.

The scenery around the third location, Okaukuejo, changed to larger trees, like eucalyptus (gum) trees, pine-like small trees, and mesquite. Some trees looked strange; Natives told us, "These trees fell from the sky." A forester called them "living petrified trees, but you should see them in the spring when they are blooming in different colours."

Swakopmund, now an international city on the Atlantic Ocean, was our next stop. We were advised not to stop in small towns on the way as young unemployed Natives roam the streets looking for opportunities to rob tourists.

We crossed the Namib Desert, the oldest desert on earth, for about two hours. We were stunned by the wandering dunes and the bare mountains in the background. After a while, quite surprisingly, the desert was covered in mist, like a fog. We entered the coastal area. Stuck in desert sand, we noticed an ancient locomotive with a sign, *"Bis here und nicht weiter"* (up to here, and not further). The meaning, we learned, was a reference to Martin Luther when he posted ninety-five theses on a church in Wittenberg, Germany, in 1517.

Suddenly, the scenery changed. A sign appeared: "Welcome—Willkommen—to Swakopmund." Fig palms lined the four-lane highway named Kaiser Wilhelm Strasse. It led us along well-treed bungalows and office buildings to the shores of the Atlantic. Side streets also carried the names of famous people during the Kaiser's time, like Bismarck and Moltke.

We found a bed and breakfast named Pension d'Agignon, operated by a French couple. It was a beautiful place with balconies surrounded by tall blooming shrubs and shade trees. We decided to stay here for a week to explore this place.

During a walk the next morning along the coast, we heard some music. As we came closer we were surprised to see a big tent with the inscription "Hamburg—Hapag Lloyd," a worldwide logistics concern based in Hamburg. The music was German. We were reminded that this was May 1, the arrival of spring. Peeking into the large tent, we saw long tables occupied by people and a band and singing groups at both ends of the tent. People waved at us to come in. We found a spot with Irish soldiers and Australians and German tourists around us. Some people stood on tables and applauded the music and singing in German as well as in English and Afrikaan. The typical bratwurst and Haake, Hamburg beer, was available. We had a good time!

On the way to Swakopmund, Namibia.

May 1st party, Swakopmund, Hansa-Lloyd tent, 1999.

A trip south to Walfish Bay, the first British harbour before the German settlement, revealed signs of shipwrecks. Here, we were greeted by thousands of flamingos, watching us as they stood on one foot in marshes along the coast. British whaling boats were visible for tourists.

Our attention was directed to a spot in the desert, about 80 kilometres inland— the location of the Welwishia plant, a species referred to as a "living petrified plant with pine-like origins!" It was 1,500 years old and survives by the mist of the Atlantic coastal air breezing inland. Each plant is protected from the steps of tourists by a fence. A packed soil would kill the plants.

Back in Swakopmund, we visited a telecom shop for e-mail and met an elderly man in the adjacent store. Hans Becker, eighty-seven years old, watched us looking at ancient carvings and overheard us speaking German. He was born in South Africa. His mother travelled to Frankfurt, in 1914 before World War I to visit relatives. They did not return until 1920 to join his father, who was interned during the war. Hans was a travelling salesman for a German import and export firm. He started his own business in Swakopmund. He and his wife live in a beautiful condo, formerly the Hohenzollern House, built in 1885. The Beckers invited us for lunch and dinner at the "Sundowner" in the Deutsches Haus and the Visitors' Club. It was an elegant restaurant in the refurbished

Railway Station, with photos and posters lining the walls. We learned a lot more about the history of tribes in Namibia.

On the way to Windhoek from Swakopmund, 1999.

On the way to Windhoek from Swakopmund, 1999.

Travelling north along the Atlantic, we drove a few hundred kilometres to get to a stretch of the Skeleton Coast. Portuguese and other European ships were thrown against the rocky coast during frequent violent storms in the thirteenth and fourteenth centuries, the time of active slave trade. Heavy road construction prevented us from getting there.

On the way back to Windhoek, we travelled through very inhospitable country by our choice. We picked up a book in

Swakopmund called *Wenn es Krieg gibt, dann gehen wir in die Wueste* (when there is war, we will go into the desert). Two German geologists did just that to avoid being interned during World War II. They published this book covering their amazing experiences living off the land in the mountains. Their studies of wildlife, fauna, forests, and desert, as well as geological studies became a masterpiece, accepted by the educational system and museums in Namibia and later in Germany. During the last of their four years in the wilderness, they were captured, but soon released to finalize their book.

In Windhoek, we stayed again at the Stolzenberg Pension to prepare for our return flight to Germany. We packaged some of our wood carvings, including a metre-high giraffe. Stolzenberg was very much interested in our travel experiences. The next morning, we had breakfast on the porch in brilliant sunshine shaded by large trees full of "diamonds" from rain the night before. It felt so good! When I commented on this to Stolzenberg, he replied, "You are right, they have a real sparkle, like the diamonds in our diamond mines south of here. If you pick up pieces outside the fences, you get shot at."

At the airport, we noticed Ursula had left her leather bag with a video camera behind. We called the lodge and were told Almuth was on her way! She was well rewarded.

What an experience for Ursula! We had travelled only through the centre and northern half of Namibia, visited many places, and met interesting people in the country in which her parents had chosen to live—the birthplace of the twin sisters!

***

Then it was back to Germany. When we arrived in Frankfurt, we leased a Bora Volkswagen Diesel (for tremendous fuel savings) for a three-week trip across Germany.

A visit with my sister Elfriede, her son Kim, and his wife Beate, in Hofheim, near Frankfurt, helped us to carry on our search for our roots.

We started at my hometown Schneverdingen, south of Hamburg. My grandfather was born in 1864 and left the homestead near Hamburg where our ancestors had built a log home and barn out of oak timber around 1712. I am sure his desire to trade goods by horse and buggy was inspired by the Hanseaten spirit of Hamburg in the Middle Ages. This era is still very much apparent in many harbour cities in Northern Europe. I sensed that trading and entrepreneurial spirit after moving to Hamburg before leaving for Canada.

We visited a distant cousin and his wife who had built a home next to the homestead. Ernst Beckedorf gave us a lot of information about the farm home, with a large thatched roof covering the barn for cattle and horses, as well. The place was fenced in and is a heritage site today. We visited a nearby historic site, a huge rock with the inscription: "Karl der Grosse 834" (Charlemagne) stopped the Saxons in their battle against his Frankish Empire at this location. A fort was built and later became a mission. Hamburg had its beginning here as Hammaburg. It grew into a major trading centre in the centuries following.

Lueneburg, the administrative centre of the region of my hometown, was worth a visit. The architecture of the timber frame homes dates back to the twelfth and thirteenth centuries. At that time, Lueneburg was the centre of the salt (white gold) trade, which extended to the Baltic Sea, the Scandinavian countries, Russia, and as far as the Middle East. The famous Lueneburg heather today covers a large area that was laid bare when pine trees were cut down to provide support for the many salt mines. The heather grew well in the bare, dry soil.

Berlin was our next destination. Ursula's relatives, her cousin Joachim and wife Annemarie, had visited us in Bragg Creek a few years earlier. We explored Berlin, witnessing the then-largest construction site in Europe, the Potsdamer Platz, and new government buildings with underground railway connections to other cities in Europe. A visit to the palaces of the Prussian royalty including Sanssouci, Frederick the Great's summer residence. It was beautifully landscaped and was worth an all-day walk.

During a previous visit to Naumburg, formerly East Germany, we visited with relatives of Ursula. The sisters of the twins' father, Adolf Buschendorf, were quite surprised to see Ursula after about forty years. Lloyd and Linda joined us at that time. A trip to Dresden and Koenigstein was a great experience for them. Moose Mountain Log Homes was re-assembling a large log home on a lake on the outskirts of Berlin. Our follow-up visit made it possible to look at the home and business of the twins' aunt. It was quite an eye-opener to see how the historic city of Naumburg had been rebuilt after the war with respect to the architecture of the twelfth century.

We left Berlin for Frankfurt and visited Erfurt, Thueringia, formerly East Germany. An active trading centre in the Middle Ages, Erfurt established the first university in Germany along with Heilbronn and Cologne. The old Gothic architecture is well maintained; the city did not suffer during the wars to the extent that other cities in Germany did. Today, Erfurt is a beautiful, green, and thriving city with many home-based businesses. The twins spent a few years in their youth in Erfurt. It was very exciting for Ursula to show me the home she lived in with Brigitte..

The twins moved from Thueringia to Wuppertal after the war. Their friends Traudel and Werner Schulz welcomed us on this trip with tours into the Rhur area, formerly the centre of the coal mining industry. This region has now become an active environmental engineering centre.

Back at my hometown, I introduced Ursula to my relatives on my mother's side, the von Fintel family near the town of Fintel, now a favourite weekend place for people from Hamburg.

Siegfried von Fintel and his family greeted us in their new home next to the farmhouse where my mother was born. I remember my grandfather when I was six years old, showing me his beekeeping operation. Looking at the farmyard, I was amazed by how well I remembered the detail of the beekeeping operation and the terrain from my visit at that time.

Honey farm, Von Fintel grandfather on mother's side, Germany, 1999.

After six weeks of travelling we were looking forward to getting back to Bragg Creek and our foothills country. I was present when Edwin, my youngest brother, passed away after a long battle with injuries he suffered after a car accident. He had been as chief chemist employed by Hoffmann & La Roche in Montreal. We brought him to Calgary after his injuries since he could not carry on his responsibilities at Hoffmann & La Roche. Edwin's death made me question life after death.

Within two months of our return from our six-week trip, another assignment was awaiting me in Germany. Previous meetings with our consultant in Nordhausen, Germany, required a follow-up visit. Ursula and Brigitte planned to join their cousin in Berlin on a trip to Prague. I visited Dr. Herms in Nordhausen for a brief review of business and took Ursula to Berlin. I returned to Nordhausen to continue our meeting, which included an engineer who arrived from Canada. The project called for a cooperation agreement with Daimler AG for the application of bio-diesel in a small delivery truck. AFS was not able to handle the challenges for costly modifications to their system.

The Harz Mountains close to Nordhausen are exceptionally pretty. The forests and meadows remind me of Bragg Creek. Earlier, Ursula and I had toured the area, including the city of Quedlinburg. It is a historic place, with a huge castle, well maintained since the time of the first German emperor was crowned here in the tenth century. The largest assembly of beautiful timber and frame homes can be seen here. Today, it is a Unesco Heritage site.

It was a coincidence that my brother Hans and Inge were in Germany for a visit with relatives. I joined Hans on a bicycle tour through the heather country near our hometown. I never tire of admiring the many beautiful places in Germany. The twins had a lot to talk about from their trip to beautiful and historic Prague when I picked them up in Berlin.

Alaska, the Mediterranean, and Italy

While in Germany, we decided to sail on the *Norwegian Sun to* Alaska with Hans and Inge. We met in Vancouver a year later. Their flight from North Carolina arrived just in time. Brigitte looked forward to joining us, but some heart challenges made it necessary for her to stay in a hospital in Vancouver.

Haida Gwaii (formerly Queen Charlotte Islands), the open Pacific Ocean, and the huge glaciers on the mainland seemed almost unreal in their beauty, overwhelming in the stark contrast of snow-capped mountains and water rushing into the vast ocean.

During our first stop in Ketchikan, Alaska, we met Ursula and Dieter Cosandier, friends from Calgary who were on a different cruise. The presence of the Russians in this town about two hundred years earlier, like Petersburg and Wrangell, is still quite evident.

The glaciers seemed to get bigger as we arrived in Skagway, Alaska. We took a train to follow the Goldrush Trail. A guide pointed out where many gold seekers were swept into a canyon in deep, wet snow with all their belongings, including horses and stoves. When the remaining caravans arrived, a terrible disappointment awaited them: there was very little gold left to find. Some people perished on the way back; others remained and eventually settled in Alaska and in bordering Yukon. I tried to put myself into the role of these gold seekers, so many people caught unprepared in a mad rush that started in San Francisco and Seattle.

On the way back, we visited Juneau, Alaska's capital, where we spent time "chewing the fat" and drinking local beer in the Gold Diggers' lounge. When we had to rush to get back to the ship, I left my jacket with my wallet behind. This was my experience of seeking gold in Alaska, the difference being that my jacket and wallet were returned to me within two months.

The travel literature from the Holland America Line got us excited to see countries with so

much history, like Italy, France, Spain and Mallorca, and Tunisia.

We met Hans and Inge in Rome to sail the western Mediterranean on the Holland America Line *M/S Noordam*. The celebration of Inge's birthday and wedding anniversaries for the four of us added a lot of excitement to the first day on the cruise ship! What a beautiful experience: sailing under deep blue skies, celebrating, and sharing our common experiences of joy. North along the Italian coast to Livorno, we went ashore to travel by bus to Florence, enjoying the beautiful north Italian scenery.

Crossing the Ligurian Sea to Monaco and Elba to Barcelona was like a dream. This city of over 1.6 million people is impressive. It considers itself at par with the greatest European cities.

The artistic explosion around 1900 is evident in the abundance of beautiful architecture. The majestic Barcelona Cathedral, surrounded by scaffolds for renovations and heavy traffic guided by police, left us with a limited view.

Palma de Mallorca was the next stop. As we strolled through the old town site, I fell in love with the unique and ancient trees with roots expanding onto the sidewalks. Resting at a café, covered by overhanging branches of huge trees, we enjoyed watching people and traffic. Ursula had some difficulty taking a photo of my futile attempts to climb a huge acacia tree. A German-speaking biologist informed us about his international work preventing the extinction of the Milan falcon, a difficult task with a population of eight hundred thousand people on Mallorca and tourists visiting the island at the rate of twelve million per year.

From here we sailed to La Goulette in Tunisia, the smallest country in North Africa. Carthage, the great African and Mediterranean metropolis from 1200 BC, was destroyed by the Romans in 146 BC. Extreme heat did not deter us from touring recent excavation sites that unearthed evidence of ancient ruins of the civilizations of Phoenicians, Etruscans before the Romans, Vandals, and Byzantines. A guided tour through Tunis led us to the mercantile section. Merchants were "hanging on to our cloth" as they tried to convince us to sample and buy their handcrafted rugs and carpets of all qualities and fibres.

Under blue skies above the Mediterranean, we sailed to Palermo, Italy. Heavy clouds and rain prevented us from going ashore. A lecturer on board gave us a history of the beautiful and ancient architecture of the city. Originally a Carthaginian colony, it was under the rule of the Normans from the ninth to twelfth centuries. Palermo became a centre of learning and an architectural mixture of Baroque and Arabic cupolas in the following centuries.

On the way to Naples, Italy, we were told of a strike by city workers. Garbage was piling up all over the city, which was already somewhat marred by the uncontrolled sprawl of industrial development.

We were fortunate to get a taxi driver who took us out of the harbour and around the city to avoid meeting piles of garbage. Once we reached the outskirts, he opened his window, leaned back, and shouted, "I love my Italy. Look at the coves, mountains, and this pristine countryside as it was two thousand years ago! Join me in singing my heart out!" Ursula and Inge joined in an operetta and his voice got louder. What a beautiful way to travel to our destination, the ruins of Pompeii! The taxi driver told us, "We meet here at this place again in four hours. Remember—or else it's a long way!"

The heat during the guided tours of the ruins was unbearable, but the sight of this historic place—the impact of what was left of Pompeii in 79 AD—made us stay on course. The former large plaestra and amphitheatre date back to the sixth century BC. The original city had a population of ten thousand, 40 percent of whom were slaves. Even though excavations continue to this day, looking at the large area still covered in ruins, I could understand why so few people survived this terrible eruption of the volcano.

Again on the way down to Naples, our driver invited the ladies to join him in singing his heart out. I, too, was inspired to join them, but afraid the driver would show me the door. Instead, I soaked in this wonderful experience of scenery and singing.

Our final day of sailing took us to Civitavecchia, where we disembarked to Rome. In full sunshine, a scenic tour through Rome offered a view of historical buildings, the Coliseum, the Arch of Constantine, various basilicas, the statue of Marcus Aurelius, and the Fountain of Piazza di Spagna. The Roman Empire came to life in my imagination.

We arrived at the airport just in time to board our flight by Lufthansa to Munich.

North Carolina and Town to Town Across Germany

We rented an Opel SUV Diesel in Munich with manual shifting. It was a great touring vehicle for our three-week trip across Germany.

Hans and Inge were acquainted with the Best Western Hotel in Erding, 20 kilometres north of the Munich Airport. Three days in this typical Bavarian city proved to be a good choice after our sea voyage.

We entered the Romantische Strasse at Donauworth, which starts at Fuessen on the Austrian border. As we travelled along this meandering, narrow, tree-lined country road, we passed many castles, ruins, and picturesque small towns. We spent the night in Dinkelsbuehl, a place first mentioned in 1200 as a secluded town surrounded by canals to protect citizens from marauding soldiers. Cars were only allowed in the evening to unload luggage in the inner city. Tourists had to walk over a hanging bridge to visit during the day.

In the evening, a traditional town crier led tourists through the narrow streets, stopping here and there for some *gluehwein*. I felt I had gone back eight hundred years in history.

Near Fulda, the Catholic *Hochburg* (dominating centre) dating back over one thousand years, we entered the autobahn toward Hannover. The Hotel Landhaus zum Alten Fritz (Frederick the Great) was under new management and did not quite meet our expectations. We walked the streets of Seesen in the evening, with people leaning out of their windows, watching us.

The next day, hungry for *bockwurst* and potato salad, we stopped close to our hometown along the autobahn. I had visited this place by bicycle in my younger years. We planned to stop here on our way back and carry on to Hamburg and Berlin. I always enjoy the trip through blooming heather country on the way to Hamburg with the beautifully timbered, white stucco farmhouses with thatched roofs, well-planned fields, meadows, and birch forests.

It took a long time to pass through the huge harbour facilities in Hamburg before we made it through the seven-kilometre-long Elbe Tunnel. The Mellingburger Schleuse Hotel was our destination. From that base, we would spend a few days wandering along the Alster River in the shade of huge oak and beech trees. We visited with relatives of Inge who lived within walking distance.

Back on the autobahn, we travelled to Neuruppin near Berlin, a Prussian garrison town with clean, wide streets. Here, we met with Dirk Froemter, a friend and former business associate, and his wife Kathrin, for an excellent lunch on the terrace of the Waldfrieden Hotel surrounded by ponds and rhododendron, lupines, and oak trees. Dirk offered us a trip through Potsdam on our way back from a visit with relatives in the northern part of the State of Brandenburg to visit Martina, Ursula's niece, 200 kilometres north of Berlin.

A three-hour trip on narrow roads was a challenge. Big trucks and buses seemed to own the road. I had to park in between large chestnut trees lining the road to let the vehicles pass.

We were guests at the eightieth birthday of Ursula's cousin Annemarie from Berlin. Her daughter Martina, the head social worker in the area, and son Rudy, a technician from Hollywood, arranged a party with many guests we knew from Berlin.

Walking through the village and along surrounding huge fields of asparagus, I was surprised to learn the experts in cutting the asparagus were from Poland. They were fast and spoke German fairly well. The local people in this rather off-the-beaten path seemed to watch us with reservation. This was a remote part of the former DDR (Deutsche Demokratische Republik), so contact with Westerners was not as yet well established.

On the way back to Berlin, we met with Dirk Froemter for a double-decker bus tour of Potsdam. This was a special opportunity to see this former showplace of Prussian royalty in its almost original condition with wide streets, well-landscaped residences, and castles.

Dirk asked us to follow him to downtown Berlin after Potsdam. This was a challenge as Dirk's Audi was hard to follow on and off the autobahn and through many sections of Berlin.

At the harbor, the site of the River Spree boat tours was very busy with tourists. We were lucky to get on. Music played and beer was served while we got a scenic tour of government buildings, museums, the Potsdamer Platz, and the Tiergarten. It was a different view than the one by car and foot we had experienced on earlier visits.

Back on the autobahn toward our hometown, we landed at the Schaeferhof Hotel on the outskirts of Schneverdingen, surrounded by sheep farms and moors. I enjoyed the environs when I walked around the hotel in the morning. The hotel had been converted at the site of an old sheep farm, with the frame house maintained with a reef-covered roof and also barns for sheep. It was a busy place with outdoor restaurant facilities. I noticed large herds of sheep on the horizon moving like white clouds. A nearby peat-harvesting operation provided wooden pathways for visitors. Signs along the way described the flora and fauna on and around the ponds, as well as the history of the now partly covered ponds from the thirteenth century to the 1950s. Dew on cotton-type plants, tall willow shrubs, and small birch trees on little islands reminded me of the beauty of diamonds everywhere. Benches allowed people to sit down and enjoy this place of peace and quiet. A sign near the entrance caught my eye. It was a quote from Hermann Loens, a famous early-twentieth-century poet : "*Lass Deine Augen offen, geschlossen Deinen Mund, und wandere still, so werden Dir die geheimen Dinge kund.*" (Leave your eyes open, your mouth closed, and wander in stillness, so secrets will be revealed.)

Friends arranged the planting of an oak tree near our hometown with a plaque that read, "Beckedorfs, Canada and USA." An area was reserved for planting oak trees for honeymooners and "sons and daughters not losing touch with their hometown." Light rain filled the air with a fresh scent. The planting of trees was a formal procedure with the guidance of the city gardener.

Former school friends invited us for lunch on their historic farm. This place, today renovated and surrounded by tall oak trees and rhododendron, had been the residence of a Baron von Zahrenhusen in the fifteenth century, a man who may be compared to Robin Hood. The name of the village I was born in, now part of Schneverdingen, is Zahrensen. The iron wood-burning stove from that time is preserved in this home.

The eighty-nine-year-old mother of my friend's wife remembered us boys scattered around the pond and corner store, then called Beckedorf's Eck (for corner). She arrived on a scooter and just plunked down on my lap! Many stories came to life, e.g., our trips to town on bicycles to go dancing and getting stuck in the mud. We washed off the shoes but had to dance in muddy pants.

Before we left our hometown, we picked up the local newspaper, which featured the oak-planting procedure with photos of us all. My niece Anne Leverenz, who lived near Schneverdingen, learned of our visit through the newspaper and called us later. She missed us by a few hours. A year later she joined us in Bragg Creek for a family reunion.

During a few days in Hofheim, close to the Frankfurt airport, we visited with my sister Elfriede, her son Kim, and wife Beate. Long walks through this town of beautiful medieval architecture are always fascinating.

A man approached us as we were sitting on a bench on a sunny afternoon in a park in downtown Hofheim. "Is this your brother from Canada?" he asked my sister. Elfriede knew him from her walks and asked him to sit with us. He sat down while feeding pigeons following him. I learned this elderly man had chosen to retire in Hofheim after spending fifteen years working for a German firm in Winnipeg and Vancouver. His wife died in Canada and he felt lonely and homesick. Two sons remained in Canada.

The return trip via Frankfurt was a special challenge at this time. The airport was undergoing a huge expansion causing detours, delays, and long line-ups. When we arrived in Bragg Creek that evening, we attended a Bragg Creek Community meeting honouring Ursula for her role in the founding of the Performing Arts and Ladies Auxiliary. A long day!

***

Our flight to Charlotte, N.C., in October 2008 took us longer than any flight to Europe. A visit here was overdue. North Carolina's climate is quite comfortable in October; also very colourful. Hans and Inge moved here about forty years ago. He was transferred from New York by BASF at that time.

During a few days in Charleston, I enjoyed the architecture of the colonial years from 1670 to 1865. A trip to Fort Sumter gave me a lot more information about the first battle of the Civil War in 1861 than I had received on a previous brief visit. After thirty-four hours of severe fighting, the Union surrendered the fort to the Confederates. The Confederates, in turn, overcame a two-year siege by Union forces. Most of the fort was reduced to brick rubble. A visit to Fort Sumter brings to life the tenacity of the Civil War and the turning point from 1861 to 1865 toward a United States of America.

The year following, we welcomed a large family reunion. Hans, Inge, and their three daughters from North Carolina and Colorado and husbands joined visitors from Germany and BC, including our relatives in Calgary. A lively crowd spent several days in Bragg Creek. We gathered at Lloyd and Linda's place, at our place by the river, and in the Seniors' Lodge. Lots of fun and sunshine!

Two years after our oak-tree-planting ceremony in Germany, it was time "to water our tree." A reminder arrived from Anne, my niece, to make sure we attended a special anniversary celebrating the blooming heather in the area around our hometown.

Hans and Inge met us in Frankfurt on a trip to Hamburg to spend a few days at our favourite place, the Mellenburger Schleuse on the river Alster. We walked under huge oak and beech trees along the river in beautiful sunshine.

The trip to Schneverdingen took us through miles of blooming heather. It was nature at its best. We stopped along the way to smell the heather! Anne and her husband Walter welcomed us to their hometown and took us to a bed and breakfast on a nearby farm surrounded by tall oak trees in a small village dating back to the Middle Ages. We walked past farm houses decorated with flower boxes and enjoyed the smell of dung along with blooming chestnut trees.

Visiting the huge and famous Walsrode Vogelpark (bird park) took a full day. We watched exciting displays of special birds: demonstrations of trained falcons, eagles, and owls along huge meadows, ponds, and ancient willow trees.

The day arrived for the annual parade and the crowning of the Heather Queen. To our great surprise, my niece, Anne, and nephew, Friedhelm, took us to an assembly of horse-drawn carriages. One of them was reserved for us!

About thirty thousand visitors lined the streets of Schneverdingen up to the destination for crowning on a high plateau overlooking the heather country. Young men and women were walking in between the carriages to make sure we followed orders and showed us how to wave at the crowd the "Queen Elizabeth way." Blooming heather along the way was still showing morning dew. "Diamonds everywhere."

People waved at us since a girl in front of our carriage displayed a sign: "Beckedorfs—Kanada und USA." The ceremony took place in a horseshoe-type canyon with a good view for the visitors.

Oak tree planting at hometown in Germany, 2008.

Heather blooming party at hometown, 2010.

After the parade, our longtime friends the Bruenings invited us to a restaurant on the Hoepen, the highest elevation around. An old sheep barn had been converted to a restaurant surrounded by porches. Many people asked us if we were homesick. My reply was, "After so many years and our children growing up in Canada and the US, we enjoy our visits to Germany very much, meeting many relatives and friends. It is a very beautiful country. We are also looking forward to getting home to the land of our choice."

British Columbia

I always enjoy travelling through the Rockies and the Okanagan by car to the West Coast, visiting with Monika in Maple Ridge. Those trips were also a means to address my impatience in early spring when I missed the smell of the earth. Monika operated a greenhouse and nursery that required keeping her land from being overtaken by salmonberries and other intrusive plants from the adjacent jungle. This was my opportunity to swing a hoe and dig deep into the rich red earth. Our great-granddaughter Marley was watching my work. She enjoyed the rides I gave her in a wheelbarrow, while Ursula and Monika enjoyed discussing the affairs of the day.

With Monika's friend Dee, we explored Malcolm Island off the northern coast of Vancouver Island. What a beautiful place off the beaten path! Finnish people had settled here around 1900 and evidence of their culture is still visible in the only town of Sointula: population 886. The island is 24 kilometres long and five kilometres wide, with scenic beaches and views of the Pacific with cruise ships passing by. Tall Douglas fir trees frame the view of the ocean with good chances for watching otters, seals, orca, and humpback whales. A little paradise.

Monika told me Dee might need some help to get rid of Salal plants in her garden. Ursula and I spent a few days on the island. I cut this evergreen plant with a scythe and then burnt the tall plants. This stuff burns easily and fast. The flames shot up into the sky—no trees nearby. Dee was happy to see her garden plot rid of the invasive plant. This was the site former settlers had used for their productive vegetable garden for many years.

White Rock, BC, is worth a visit. The view across the Pacific Ocean to the U.S. makes this resort special. The tree-lined avenue along the coast offers neat little cafés and restaurants and a long beach filled with families enjoying the bordering oceanfront . It is a relaxing place to watch people meandering along the avenue and beach. I pointed out the beautiful young trees planted along the boulevard to the owner of the small café. He shrugged it off. "I don't like the trees. They leave too many leaves." I suggested Mugho pines would enhance the beauty and not interfere with the view of the boulevard and the ocean. "Okay, let's make a deal." A tempting situation.

Our favourite "halfway" place to the West Coast is West Kelowna. Visiting Brigitte and Herbert was always a good opportunity for Ursula to spend some time with her twin sister. Hiking to the Trestle Bridges built by the CPR around 1900 in the Kettle Valley, Herbert and I explored the country where we worked in a portable sawmill in 1952. The road to the sawmill was in the same poor shape as it had been fifty years earlier.

Ursula, Cressen, Marley in Vancouver, BC, 1992.

Sometimes we chose Highway 3 through the Kootenay for the return to Bragg Creek. While driving along Kootenay Lake, a story came alive. The father of a rancher, our neighbour next to our first weekend place between Bragg Creek and Priddis, lost his belongings in the lake. Heather Crawford told us her father came to Canada from England to buy a ranch in the foothills of Alberta. He had travelled from Vancouver and, while crossing Kootenay Lake, his canoe had capsized in stormy weather. Saving himself by hanging onto the canoe, he lost his wallet with all his cash and some other belongings in the lake. He worked on Alberta farms until he could afford to buy the land where his daughter is still ranching.

In the meantime, my brother Ewald and his wife Lilo had followed our trails "into the sticks" when they acquired an acreage in the Sundre area. Our visits to their place made us aware of the same enthusiasm they expressed in their love of the foothills country.

Cochrane

This active town developed into one of the fastest-growing places in Alberta without losing its Western ranching atmosphere. All services are now available in Cochrane.

What a change from 1953 when we arrived there from BC on Highway 1A and the only business establishments were the Cochrane Hotel, a restaurant, a service station, a railway connection, a few homes, and very few trees. The view of the foothills and mountains is amazing. We looked down into the valley from the Cochrane Hill at a large and peaceful-looking sheep farm. This valley is now covered by the sprawling GlenEagles subdivision.

Our land acquisition in West Bragg Creek was negotiated and settled in the Cochrane Hotel. Lloyd and Linda established Moose Mountain Log Homes in Cochrane. Cyr, our grandson, has attended school from the beginning here. Linda`s sister Jo-Anne Bourdage, with husband and son Wesley and Daniel, made their home in Cochrane. Here, we meet often with Linda`s parents, Celine and John-Guy, their daughter Manon Bourdage, and husband Will for family reunions . Frank Schroeder, our nephew, and his wife, Patty, have lived in Cochrane for many years and operate Glenbow Electric there. It is great to watch tree-planting efforts in the area, adding to the beauty of Cochrane.

♦

THE GREEN GRASS OF HOME

It is always a great feeling to come home to the Alberta foothills, the land of the four strong winds that blow lonely."

Bragg Creek

Green grass and trees are good for the eyes. Waiting for the grass to turn green takes patience in our area. When our daughter Monika phones from Maple Ridge and "brags" about the flowering world around her in March and April, I reply, half-seriously, "Oh, yeah? At least we don't have to start cutting the grass and digging in the garden as yet. We take it easy for another two months."

The hamlet of Bragg Creek had its beginning as a community of ranchers and a weekend place for Calgarians. The growth of Calgary increased the population of the hamlet and clean drinking water was at risk. A Bragg Creek Water Committee was formed in 1999 as a liaison between the citizens of Bragg Creek and the municipality of Rocky View. The committee worked over a period of two years to assist in finding ways to bring clean water to households at a reasonable cost. As a member of this committee, I was disappointed in the results. Provincial and municipal authorities were not able to follow through with their commitments for funding after feasibility studies by two engineering firms were submitted.

At this time, a group of seniors under the leadership of Terry Graham assembled a plan to build a seniors' lodge on land leased from the MD of Rocky View. Moose Mountain Log Homes assisted in the construction of a beautiful log structure for this purpose. Lloyd and Linda arranged to have two log sections brought in from a log building school in Kelowna and filled in the entire log structure.

This lodge was completed in 2000 with the help of seniors and funding matched by agencies of the provincial government. Groups renting this facility called it "the best seniors' lodge in Alberta." The landscaping of the property was enhanced by a dozen Colorado blue spruce trees provided by Ursula and myself. Today, these trees already fulfill the purpose of privacy.

The Bragg Creek Senior Housing Society has a mission "to create a community-based Supportive Living framework for area seniors, which includes affordable, accessible, and ecologically sustainable needs-driven programs and services." I joined this group as a director in 2009. We are making progress; the challenges are great.

When my sister Irmgard suddenly passed away in 2007, I became more aware of the importance of enjoying each day to the fullest of our ability. The celebration of her life at Lloyd and Linda's place brought together another large reunion of family and friends.

The Young Explorer

Sitting on our porch, we felt fortunate for having travelled to many countries and been able to return to the stillness of the forest and beauty of the foothills, enjoying the best of both worlds. Our tree farm of Colorado blue spruce welcomed us with a display of diamonds in all colours!

A family reunion in 2009 showed the original clan had expanded to a large group, now assembled at our place with more than fifty people of all ages. Ewald's son Tom, his wife Brenda, and son and daughter, Mitchel and Aubrey, had joined the clan.

We celebrated Cressen's daughter Marley's birthday in 1992 in Vancouver. Monika, the proud grandmother, looked so young and happy. Simon's happy face and the great-grandparents expressing their joy contributed to a very happy reunion.

Ten years later, we were shocked when Cressen died suddenly of cancer. A family reunion for the funeral brought us together to support Monika and Simon in these difficult times.

Cyr, Linda and Lloyd's son, our first grandson, was born on January 2, 2000. Ursula and I enjoyed that lively young explorer. When Cyr was one year old, we joined his family for a trip to Hualtuco, Mexico, and witnessed his first swimming exercises. Soon, he became an excellent swimmer. After a return from a holiday in Yuma, Arizona, we visited Linda and Lloyd and one-year, one-month, one-week, and one-day old Cyr. This occasion was his first upright walk, when he tried to greet us. His parents often dropped him off at our place on the way to work. We had the fun and the fortune of feeding him, walking with him, and exploring our surroundings in his company. Schatzi, our dachshund, and I were his steady companions. Cyr researched the creepy crawlers in the neighbour's pond. I watched him holding a stick to fetch interesting things; he studied his catch and shared his research with me. He held hay to the horses' mouths, talking to them constantly.

Cyr feeding horses, Beck' N' Busch No. 4, 2003.

Cyr the researcher and Schatzi, our dachshund, 2003.

Cyr, full of excitement, at our log home, 2004.

While I was recovering from back surgery, three-year-old Cyr visited me with his parents and others. He came to my bed. I asked him to tell the others that I was in pain and couldn't talk. With a determined posture and waving his hands, he told everybody, "Opa can't talk!"

Korina and Cyr at our river frontage, June 2007.

Cyr the explorer of geological formation, at our river frontage, 2009.

Cyr and cousin Daniel taking a rest at our river frontage, 2010.

Louie, a Golden Doodle puppy, joined the family when Cyr was seven years old. Louie was born on the same day as Cyr. Racing Louie up the hill to the log home became a favourite sport. Whenever the family travels on holidays, Louie spends the time with us on the river. He is a real family member, loved by all.

A Dream Log Home

Lloyd and Linda spent many years building their own log home on a beautiful acreage south of Bragg Creek. Their love of the art of building unique log homes was reflected in their success in many parts of the world. Their site provides a panoramic view of the mountains, the foothills, and the meadows.

As the building took shape, Ursula and I worked out plans with Lloyd and Linda to move into the large home. The idea was to downsize from our own log home to a suite on the walkout floor under construction. We looked forward to being able to travel to places not yet visited.

There was a lot of work in store for us to move. Stuff had accumulated over twenty-five years and much of it was no longer needed. Garage sales, donations, and "dumping" were the answer. We put up a "For Sale by Owner" sign and sold our place within a short time in the summer of 2006 with a condition to remain in the home until we were able to move. Vera Krueger, the new owner, added to the beauty of the home and surroundings. Recently, she converted the property into a guest home, a favourite place for relatives and friends from Germany.

What a shock on Remembrance Day, November 11, 2006. A dramatic turn of events took place—Linda, Lloyd, and Cyr were ready to move into the new home by the end of November. A natural-gas-powered electrical generator caused a fire in the garage, where a lot of building material was located. The gas plant was well maintained and working for many years during construction. When Lloyd arrived in the morning he noticed some smoke coming out of the garage. He immediately called the local fire department. Some members were participating in the Bragg Creek parade, causing a delay. Precious time was lost. Even with the help of a neighbour's pump connected to a water cistern containing three thousand gallons of water, it was too late. The fire consumed the building materials stored in the garage before it spread to the house.

When I arrived, the home was engulfed in flames. I stood next to Lloyd watching the floors cave in. He said in a calm voice, "It will take some time to rebuild." Linda could not believe her eyes. She was crying and leaning on Lloyd. Years of hard work gone, just like that.

Both Lloyd and Linda were determined to rebuild as soon as possible. With the help of a track-hoe, Lloyd was able to clean and level the building site. Some poplars were lost, but almost fifty Colorado blue spruce trees, planted earlier, were saved. The new building site allows for more space for equipment and materials to be moved around the new home. The revised plan offers an even greater mountain view, a beautiful new log home creation in progress.

Ursula and I, as well as Linda, Lloyd, and Cyr, had to find new homes. In January 2007, all of us settled into new surroundings within the hamlet of Bragg Creek.

A property on the banks of the Elbow River became available. Even though the ground was covered in snow, Ursula and I were excited. The south-facing river frontage intrigued us; the open spaces and terraced frontage to the river offered opportunities for outdoor enjoyment and entertainment. The oversized double garage had a lot of storage space, which had the disadvantage of allowing us to keep stuff we no longer needed. The lower floor of the bungalow was developed and the house offered more space than we were looking for. The best part was the view of the river. The large frontage allowed for a lot of improvements to suit our outdoor needs and to take full advantage of the southern exposure. A two-bedroom suite on top of the garage provided additional income and tenants to look after our place when we travel.

I enjoy landscaping, building rock gardens, and planting Mugho pines and a variety of shrubs. I "planted" an old wagon wheel with part of an axle attached leaning on a tree. This wheel reminds me of an amazing technology invented by our ancestors to move stuff around. It also demonstrates to me the cycle of nature—our lives and the four seasons. I added an antique and idle water pump—useless, but fitting in. I consider my rock garden a work of art, placing colours in ways of harmony and beauty, allowing each plant its growing space, adapting to the change in seasons.

Family reunions fill our "beach" with a lot of activities. BBQs around open fires, a pit protected by large fieldstone rocks with Ursula and friends singing and music are fun for everyone. Seeing Cyr and cousin Daniel and friends jumping from rock to rock in the shallow river followed by Louie, Cyr's dog, and finding fossils along the embankments, is great to watch.

The Elbow River

Elbow River frontage in snow.

Landscaping the river frontage.

The rock garden, 2014.

The Elbow River

The year-round beauty of the "Elbow" has fascinated me in many ways. It changes its course often, moving gravel banks and revising the scenery in the process. We are fortunate to have at our frontage ancient shale meeting sandstone, gravel and large boulders, creating rapids. The waves jumping and spraying billions of drops of water heading for the lowest point in the watershed make me wonder: How many billions of molecules in each drop?

Special Internet programs such as Oasis and Deutsche Welle TV programs feature interesting points about the origins of water. What if comets hadn't brought water and minerals to our planet during the first billion years after the Big Bang? Probably there would be no life on earth! The present exploration of a comet may bring us a clue as to where in the universe water originated.

The sound of the waves rushing over the boulders is music to our ears. The time spent, summer and winter, by the river is very precious to Ursula and me. It is a good place to meditate, expanding awareness of an inner spark of energy flowing from head to toe. I let it flow.

The smell of trees and flowers and the glittering of diamonds on the grass and shrubs by the river, as well as the stillness in the wintertime, calm my mind. I feel the energy of nature in every part of my body whenever I take the time to sit down and relax by the river, feeling the rhythm of nature.

Willows by the Elbow River in the fall, 2013.

Open fire pit picnic area by the Elbow River, 2009.

Playing Bridge by the river is an opportunity to share our place with neighbours and friends. Jean and Jack Leslie visited us often to play bridge. Jack, the former mayor of Calgary, told us, "I was born on the banks of the Elbow River in Calgary. The view from here is one of the best I have seen."

A Life of Song and Music

After Brigitte, Herb and family moved to Kelowna, Dona and Ross Mutton visit often. Playing guitar, singing with Ursula, and playing bridge are ongoing activities we maintain by visiting them in Calgary regularly and having them join us by the river. I enjoy music and singing, the sounds filling the beautiful scenery surrounding us. I thank the twins for their talents in playing the guitar and singing for many years, adding to my appreciation of music.

◆

THE FLOOD OF 2013

The Power of Gravity

The flood of June 2013 was the largest in provincial history according to *Maclean's* magazine (July 15, 2013). Massive rains, melting snow in the mountains, and an unusually cool spring caused rivers to overflow their banks. High and tumultuous waves brought havoc to southern Alberta, especially High River's population of thirteen thousand. Bragg Creek residents experienced substantial damage. The commercial area was devastated. The forest floor in Bragg Creek and area filled with piles of mud. There were evergreens scattered everywhere.

We could not believe our eyes after the first onslaught of huge waves: a meadow (at least two to three hundred years in growth along the river front) was swept off its shale base with roots of grass, silver-willows, alder, and poplar trees rushing downstream, followed by rooftops and a twelve-by-three-foot metal well casing sailing by. It took a while for us to get used to an amazing new view: a "patio of shale" in front of our riverbank, the bridge in Bragg Creek across the Elbow River in full view of our living room window. The flood gave the river "elbow freedom" to uproot trees on both sides and change the river frontage, widening the river valley considerably. How fortunate we were to have our landscaped lawn area untouched by the river and no flooding in our home!

We were surprised to see an amazing number of volunteers arriving on the scene. This great effort of helping neighbours and friends clean up their places was organized by the Bragg Creek community.

A Young Life of Love and Talents

Within twenty-four hours of the Big Flood reaching its peak, it became evident that many places in Bragg Creek were severely damaged. Our daughter's place in the centre of the hamlet was among the worst hit. On June 22, Korina's birthday, we learned she had been evacuated with other flood victims.

We reached Korina the same day in Bearspaw, where she was staying with a couple. Gracie, her dog, was with her. We wished her a happy birthday! She told us the couple took her to a dog show with their own dog. "I am at the end of my strength; my stuff under the tarp is spoiled. The cabin is flooded."

We were fortunate. The flood pushed debris and sand across our river frontage but no water entered our home. Five relatives from Germany, including Anne, my niece, and Friedhelm, my cousin, arrived two days before the flood to visit us and other relatives in BC and North Carolina. We enjoyed a sunny day along the riverbank before the flood. It was their plan to travel through the mountains on the day the flood hit. In the meantime, the bridge across the Elbow River was closed. Our visitors were stuck in West Bragg Creek in a bed and breakfast (our former log home). Our place offered a good opportunity for sightseeing, so they spent a couple of days with us and helped to clean up our river frontage. They caught the roaring noise of the three-and-a-half-metre waves on video.

Provincial government authorities for disaster-relief funding told Korina she could not move back into her home until an assessment of the damage and contamination of the water supply had been completed. She had to clean up the inside of her home before the assessment could be made. She put most of her belongings outside on her lawn and covered them with a large tarp. A neighbour took Korina's beautiful Irish wolfhound while she stayed in temporary accommodations in Cochrane.

Late in August, Korina moved back into her home to help speed up the cleaning process. Constant rains delayed the cleaning crew and a leaky tarp damaged more of her belongings outside on the lawn. Korina's health deteriorated. She passed away in her sleep in her home on August 29. The sudden loss of our daughter was a severe shock to us all. We are struggling to accept the cruel facts.

On the day of the Celebration of Korina's Life, family and friends joined us early in the morning on a traditional trip to a beautiful spot on the Moose Mountain meadows, Korina's favourite place. We met here also for other family members who had died earlier. Lloyd spread her ashes in brilliant sunshine with Moose Mountain in full view. We shared many good memories.

Linda organized a large gathering in the afternoon of relatives, friends, and neighbours who joined us in the Celebration of Korina's Life in Redwood Hall. Heartfelt comments and speeches emphasized Korina's love of nature, her gentle attitude and willingness to help other people, her involvement in keeping Bragg Creek green, and her love for her dogs. We celebrated a beautiful person with many talents.

A feeling of emptiness, missing Korina, put me in a trance. Busy days followed to take care of the effects of Korina's death. Ursula and I consoled each other with reflections of good memories of Korina's life.

♦

HEALTH CHALLENGES

Our days in the "wilderness" were exciting and at times challenging. A heart condition forced Brigitte to spend time in Edmonton Hospital in 1971 and later in Vancouver.

My brother Hans and his wife had a very active life. Ursula and I traveled a lot with them on cruises into the Mediterranean and up to Alaska, and several trips to Germany. It was great fun traveling with them while we enjoyed life in the moment. Now they experience health challenges and are losing their independence. We are thinking of them a lot.

Ursula's heart operation took place in January 2013 in the Calgary Foothills Hospital. The recovery process was very slow. The heart specialists are recognized as some of the best in the field. We were told patience and exercise are the best medicine. Another heart operation was not recommended, and special medication was prescribed. Frequent visits to cardiologists and other specialists were necessary. Gradually, Ursula's condition deteriorated. We were fortunate that, within one year of the operation, the Care in the Creek medical clinic opened up in Bragg Creek and home care by Alberta Health Cochrane was made available to us. Ursula and I spent a lot of time by the river frontage, walking and enjoying the fresh air. All of this made life fairly comfortable for Ursula. We celebrated our 60th wedding anniversary in May 2015 with blue skies and the sound of the river.

A Life of Song and Music Coming to an End

Ursula's condition became very critical.

I Want to Die in my Home

"I do not want to go to the hospital, I want to die in my home." Ursula's special request had to be accepted by the professionals. She died on September 14 in our home with the family close by. It was a deep feeling of love, very peaceful for me to comfort Ursula with cool and wet towels. When Ursula raised her head and opened her eyes for the last time, I kissed her forehead, feeling the loss but also relief. Her cold body was taken care of by cremation specialists later the same day. When the body passed by me, I knew it was not Ursula's body anymore. She was very much present with me in all phases of my life and those of my family.

A Celebration of Ursula's Life in the Seniors' Chalet in Bragg Creek on October 18, 2015, was attended by a great number of relatives, friends, and members of the community. This was no surprise to our family and friends considering Ursula's

contributions in building an active, artistic, and resourceful community. A video playing in the foreground showed Ursula in almost everyone of the about two hundred photos smiling or singing with guitar in hand. A Life of Song and Music was well celebrated.

Catching My Breath

When I look at a tree, I see a living entity thriving for sunlight. One day, its strength will give in and it will fall down and nourish another tree. Nature is full of energy, immense power, beauty, and mysteries. We are part of nature. Being conscious of our source puts life into this energy. I let it flow from head to toe, knowing my inner strength, love, and freedom.

ABOUT THE AUTHOR

With four brothers and two sisters, Siegfried grew up in a small village south of Hamburg, Germany. Their outdoor activities and explorations of the beautiful surrounding forest, ponds, and meadows was supervised by a maid playing the guitar. Their mother died shortly before World War II when Siegfried was nine years old. The war years left memories of violence, devastation, and death, but also many encouragements like "diamonds everywhere."

With an urge to see the world, Siegfried and two brothers sailed to Canada in 1951 for a two-year adventure, not realizing they may not come back in that time frame. Their travels led them to Calgary at the time of the Calgary Stampede. At a dance, Siegfried met Ursula and her twin, Brigitte. Ursula was a beautiful dancer and instructor at the Arthur Murray School of Dance. While dancing, she sang along with the band. Siegfried was in awe. He forgot all about returning to Germany as planned. They got married in 1955, settled down in Calgary, and later followed a dream to move to Bragg Creek in the foothills west of Calgary. Brigitte married Siegfried's brother, Herbert. The Beckedorf Twins entertained with guitar and song at many festivities in Calgary and in the Bragg Creek area. While Ursula participated substantially in building an artistic and resourceful community in Bragg Creek, Siegfried got involved in land development and international business.

Ursula and Siegfried traveled frequently. The highlight was a flight to Namibia, Southwest Africa, the twins' place of birth. The couple celebrated their sixtieth wedding anniversary, and the fiftieth anniversary of their having been residents of the Bragg Creek area, in 2015. A life of song and music came to an end when Ursula died in their home in September 2015.

CPSIA information can be obtained
at www.ICGtesting.com
Printed in the USA
LVOW05s2341050517
533454LV00007B/9/P